GRIZZLY
CONFIDENTIAL

Also by Kevin Grange

Wild Rescues: A Paramedic's Extreme Adventures in Yosemite, Yellowstone, and Grand Teton

Lights and Sirens: The Education of a Paramedic

Beneath Blossom Rain: Discovering Bhutan on the Toughest Trek in the World

GRIZZLY

CONFIDENTIAL

An Astounding Journey into *the* Secret Life
of North America's Most Fearsome Predator

KEVIN GRANGE

HARPER
HORIZON

Published by Harper Horizon, an imprint of HarperCollins Focus LLC.

Any internet addresses, phone numbers, or company or product information printed in this book are offered as a resource and are not intended in any way to be or to imply an endorsement by Harper Horizon, nor does Harper Horizon vouch for the existence, content, or services of these sites, phone numbers, companies, or products beyond the life of this book.

Interior design: Kait Lamphere

ISBN 978-1-4003-3826-9 (eBook)
ISBN 978-1-4003-3825-2 (HC)

Library of Congress Control Number: 2024932549

Printed in the United States of America

24 25 26 27 28 LBC 5 4 3 2 1

For those who love living with wild things

Contents

Author's Note

Ursus arctos commonly refers to all brown bears, of which there are two subspecies: *Ursus arctos horribilis* and the great bear of the Kodiak Archipelago, *Ursus arctos middendorffi.* Historically, the term *brown bear* referred to bears that had access to the ocean and marine life and *grizzly bear* was reserved for bruins found in the interior of North America. However, since the genetic difference between these bears is negligible, I use the terms *grizzly* and *brown bear* interchangeably throughout this narrative.

The research for this book took place over two years at the tail end of COVID. I conducted interviews both virtually and in person and have retained the content of these exchanges during the days I spent in the field with my subjects, though in some cases I've changed identifying details to protect individuals' privacy, condensed multiple days in the field, or generalized dates, times, and order of events. I've also edited some quotes for length and clarity. The events and conversations depicted are based on my own recollection, and I've worked hard to ensure the details are accurate.

Kevin Grange
Jackson, Wyoming

What is man without the beasts? If all the beasts were gone, man would die from a great loneliness of spirit. For whatever happens to the beasts, soon happens to the man.

—Chief Seattle

What most people know about them still has less to do with the nature of grizzlies, than with the nature of stories.

—Douglas Chadwick

Walking Among Giants

I think I could turn and live with animals; they are so placid and
self-contained. I stand and look at them long and long.

—*Walt Whitman*

As our flat-bottomed boat left the bay and motored upriver, I reached into the backpack at my feet and pulled out my binoculars. Our guide referred to the area as "the nursery" because it was teeming with mama brown bears and cubs. Gazing through my binos, I counted three different sets of adult females and their offspring wandering in the surrounding tidal flats.

The sows, who had the exhausted look of all new mothers, were busy digging for fish eggs in dark gravel, chasing silver salmon in shallow water, and munching on tender tidal grasses. For months these devoted mothers had been teaching their little ones not only what to eat but where to find it and, most importantly, when the food source is available.

But the spring cubs were too distracted to pay close attention. They were busy being siblings, which, that morning, meant wrestling matches, scrambling up the crooked arms of driftwood, and playing keep-away with half-eaten fish carcasses.

I'd traveled to Kodiak, Alaska, with my wife, Meaghan, to see the largest brown bears on earth, and the archipelago didn't disappoint.

The island was crawling with bears—hulking bruins whose bellies nearly dragged on the dirt and who ambled with wide, prehistoric gaits.

Suddenly, we heard steel grinding over river rocks. Our guide, Hiram, cut the motor.

"You'll have to walk from here," he announced gruffly to us and our fellow passengers. "The tide's too low."

Like many in the far north of Alaska, Hiram had held many jobs over the years—commercial fisherman, carpenter, bouncer, and construction worker—but it was clear that what he loved most was being a fishing, bear-viewing, and hunting guide. With his worn baseball cap protecting a head of unruly gray hair, his wide muttonchops, his chest waders, and his long trench coat—held together in places with silver duct tape—Hiram appeared to have stepped from the pages of a Farley Mowat novel. He had a voice like God in the Old Testament, and his instructions to me and Meaghan on the correct way to spin fish were less suggestions and more like holy commandments: *Hold the rod in your dominant hand. Stick your index finger out and pull the line in. Flip the bail on the spool to unlock the line!*

Hiram also had an inspiring and flawless environmental ethic when it came to the grizzly bears. If we were motoring upriver and saw a bear in the water, he'd cut the engine a hundred yards away so we wouldn't displace the bruin. If we were reeling in a fish and a bear approached, hoping we'd run away and leave him dinner, Hiram would make us cut our line so the bear wouldn't associate humans with food. For the same reason, he'd wait to clean fish so as not to leave entrails along the shoreline.

As my wife and I and our fellow passengers—two middle-aged blokes from London who, like us, had come to fish and watch bears—exited the skiff, Hiram announced, "I'll meet you up ahead where the river bends and water deepens."

On other occasions over the past three days when we'd similarly disembarked, Hiram had kept pace with us, walking while trailing the boat on a rope as if leading a horse—but today he hopped back in the skiff, hit the motor, and took off.

"There goes our bear-defense rifle," I remarked.

"Guess we're on our own," said Meaghan, laughing.

We took our time, pausing to adjust our backpacks and drink water, allowing the other passengers to walk ahead.

"Can't believe we leave in twenty-four hours," I said, after having spent four days at camp.

"Yes," Meaghan replied, "let's make it count."

The tidal flats were carpeted in sedge, sand, and decaying fish carcasses. Like many rivers in Kodiak, this one had multiple salmon runs. First sockeye and then pink, chum, and silver salmon, creating monthslong buffets for the bears, seagulls, and bald eagles. Thanks to the low tide, I'd come to savor these daily strolls. The walk off the skiff each morning was a kind of threshold crossing, from the human habitation back at the bear camp where we were staying to the wild land we were entering. For those few moments, my anxious monkey mind calmed. I felt fully present and filled with gratitude for family, friends, good health, and, of course, my good fortune in being about to spend the day bear viewing in Alaska.

Just then two cubs spilled out from the sedge, five yards to our right. The coys (cubs of the year) were adorable with honey-colored fur, fluffy ears, and white natal rings around their necks.

But Meaghan and I had a more pressing thought: *Where's mom?*

A large sow suddenly reared up in the tall grass behind the youngsters, dropping a fish from her mighty jaws and fixing her eyes on us.

It was a surprise encounter with a sow and cubs, at close range, by a food source. If one was to pick a "most likely to get mauled by a brown bear" scenario, this would be it.

"Not good," I whispered to Meaghan.

My sympathetic nervous system immediately kicked into overdrive. Adrenaline and cortisol fizzed through my veins. My heart jackhammered in my chest, and my breath came in rapid, shallow bursts. *You're going to die—run!* screamed one part of my brain, while the other parts told me to stay cool and calm.

"Respect the bear and the bear will respect you," the Elders of Kodiak have preached for millennia.

Like a lot of Indigenous wisdom, the mantra seemed so simple and linear when I first heard it. But the more I thought about it, the more complex, cyclical, and way beyond my comprehension it became.

Grizzly bears can run thirty-five miles per hour, crush a bowling ball with one bite, and kill a massive elk with a sudden swipe of a mighty paw, so I knew that neither running nor battling the bruin was an option.

Torn between fight or flight that morning, Meaghan and I chose a third option . . . we froze. To our relief, the bear didn't attack. To be honest, I hadn't really expected her to. My journey into the secret life of brown bears had already taught me that everything I'd thought I knew about them was wrong.

Grizzlies aren't ferocious man-killers, nor are they cuddly teddy bears. Instead, they inhabit the fascinating landscape between these two extremes. I'd always assumed they were primarily carnivores—in truth, they prefer a mostly vegetarian diet with the occasional side of meat or salmon. Brown bears are unpredictable? Turned out that's also incorrect. Brown bears express *exactly* how they feel using posture and vocalizations, like jaw popping or huffing. Grizzlies don't so much defend a territory as have home ranges. And they aren't true hibernators either.

"They hibernate," I insisted to one biologist. "They enter the den in the fall and sleep until spring. You can't refute that!"

She informed me that their hibernation was more like a state of torpor. "It's actually a series of sleep-wake cycles," she explained. "The body temperature drops only a few degrees, and they can arouse almost instantly if they perceive a threat."

I'd thought grizzlies were solitary, but it turns out that they don't mind sharing space with other bears if there's an abundant food source, and from time to time, siblings or family groups will pal around together. Big old boars don't always kill young bears—in fact, they occasionally hang out with them. Mother bears sometimes adopt orphaned cubs, and sows with cubs don't always attack during surprise encounters with a human at close range. There were many other behaviors they could, and did, choose when encountering people unexpectedly.

I also discovered that brown bears have superpowers. They can "smell time" and "see smells," as one biologist described it. Grizzlies are great adaptors. They've outlasted the woolly mammoth, the saber-toothed cat, and the mastodon. They can live on high mountaintops, in the forests, the Arctic tundra, subalpine meadows, or desert. Brown bears can gain hundreds of pounds each year yet not get diabetes or heart disease. They can sleep for months[1] while their heart rate drops as low as ten beats per minute, yet, despite being immobile, somehow lose little muscle mass and don't develop blood clots. And during those months when they're not eating or drinking, their bodies can actually recycle—and reuse—waste products. Pregnancy? That involves a process called delayed implantation, where the fertilized egg doesn't implant into the uterus right away but waits in suspended animation for months until the sow gains enough weight to support her pregnancy. Once the blastocyst finally implants, the mom gives birth just seven to nine weeks later, while she's still hibernating.

Despite their size, strength, and superpowers, brown bears are still surprisingly vulnerable. They, too, have kryptonite: humans—fellow predators, bipeds, and omnivores who compete for the same land and resources and hold the fate of the grizzlies' survival in their hands.

Regardless of knowing that a sow with cubs won't always attack, I was still terrified during the close encounter Meaghan and I had that morning. I carried bear spray with me but had committed the cardinal mistake—it wasn't accessible and ready to deploy within three seconds. Instead, the useless can was wedged into the front pocket of

---

1. Jack Tamisiea, "How Bears Hibernate without Getting Blood Clots, *Scientific American,* April 13, 2023.

my sweatshirt, accessible only by unfastening two buckles and pulling down my chest waders.

"Hey bear," I said in a soft but firm voice.

Standing tall and sniffing the coastal winds, the sow glanced around before dropping back down on all fours and picking up the fish.

"Let's back up slowly," Meaghan whispered.

The two of us walked on, glancing over our shoulders until we finally reached Hiram and the boat. As we did, I felt a trigger surge of elation. By all accounts, our bear encounter had been a resounding success: we hadn't stressed the family unit or changed their behavior. And, most importantly, we hadn't been charged.

"I've been watching wildlife for forty years, and that experience beats all the others," I exclaimed.

My passion for watching wildlife began when I was twelve years old and my parents gave me a new edition of the classic *Field Guide to Animal Tracks*, written by the great biologist Olaus J. Murie in 1954.

To Kevin, our family's great pathfinder, fisherman, and outdoorsman on his birthday, my father inscribed on the title page.

At the time, my family lived on forty-six wooded acres in Brentwood, New Hampshire. I was in sixth grade and reading all the great books about boys and their dogs: *Old Yeller, Sounder, Shiloh*, and *Big Red*. I dreamed of hunting racoons with two redbone hounds like Billy in *Where the Red Fern Grows*, but instead of two new puppies, I shared ownership with my siblings of a mutt named McGee. As for hunting, no one in my family had any experience, so instead of a rifle, I received the *Field Guide to Animal Tracks*, a subscription to *Field & Stream*, and a Havahart metal cage trap with two spring-loaded doors that promised to safely and securely catch racoons, rabbits, and squirrels.

Far from being disappointed with my live animal trap and our timid beagle–black Lab mix, I was thrilled, wandering the woods with my guidebook and scouting for animals. During my backyard adventures, I spotted whitetail deer, racoons, squirrels, skunks, and red foxes and looked—futilely—for bears, the true emblems of the wilderness.

I was fascinated with bears. And as I looked around my life, I noticed bears everywhere. A teddy bear tucked me in each night, promising companionship, comfort, protection, and sweet dreams; my mom, a loyal and protective mother bear, gave me a bear hug before I left for basketball camp each summer; my father lamented the drop in stocks that occurred with a bear market; Smokey Bear—with his flat-brimmed ranger hat—reminded me "Only You Can Prevent Wildfires"; in geography class, I spotted a grizzly bear prowling prominently across the California state flag; and on a starry night, I would gaze up at the constellation Ursa Major (the Great Bear) in the northern sky with the utmost wonder.

I pored over the bear prints in the *Field Guide to Animal Tracks*. The bean-shaped foot pad and oval toes were strangely human-like, and there was an enchanting calligraphy in the claws.

Hoping to see them, I memorized the difference between *Ursus americanus* and *Ursus arctos*: the claws of a black bear are smaller, and the toes are more separated and curved. Grizzly bear claws are larger, and the toes are closer together and less curved. But the ultimate test involved employing a ruler to find the lowest point of the outside (largest toe) of the print. From there, I'd find the highest point (or front) edge of the palm pad. Connect the two points with the ruler and if 50 percent of the inside (smallest) toe is above the line, it's a grizzly bear. If more than 50 percent of the inside (smallest) toe is below the line, it's a black bear.

I never stumbled across any bruin tracks in my backyard, but my family did run into a black bear later that summer during a camping trip to Moose Brook State Park in the White Mountains. We'd gone for a night hike around the campground after dinner, and as we walked, I sensed that a bear was nearby. Like some kind of animal ESP, I could *feel* the bruin before I saw him, and as we approached an old water pump beside a lilac bush, I knew he was hiding in the dark shadows.

Suddenly, we heard a low growl that clearly meant, "Don't come closer."

"Bear!" I yelled, and my family did the one thing you should never do when encountering a bruin: we ran and, by doing so, mimicked prey. Our mutt, McGee, definitely wasn't from the Old Yeller lineage—he left us all high and dry, clinging to the old adage "You don't have to outrun the bear, you just have to run faster than your friends."

Thankfully, the bear didn't chase us. But shadowy ursid apparitions ambled through my unconscious in the nights that followed, causing me to wake up and look under my bed. The bear had become more present because of its absence.

In between homework assignments, I read more about bears and discovered there were eight species in the world: Asiatic black bears, pandas, sun bears, speckled bears, sloth bears, brown bears, black bears, and polar bears.

Grizzlies, black bears, and polar bears are all found in North America, though polar bears are found only in Alaska and the northern-most parts of Canada. Despite their names, color is not a great identifier for bears. There are "brown" black bears, "black" brown bears, and every shade in between—cinnamon, blond, bluish-gray, and even white. The Kermode bear, also known as the "spirit bear," is a white subspecies of black bears found in the central and north coast of British Columbia, and there are rare cases of white grizzly bears, such as one near Banff, Alberta, that locals called Nakoda, named after the Indigenous people of Western Canada and the United States.

I discovered the best way to differentiate between black and grizzly bears is by their size and shape rather than their fur color. Brown bears are generally bigger and can be distinguished from black bears by their long claws, dish-shaped face, and distinct shoulder humps—a large mass of muscle used for digging and finding food sources.

Later during that summer of my twelfth year, I saw my first black bear, but it was hard to call it "wild" since it was feeding on garbage in an open-air dump with a bunch of other hungry bruins. That was in the 1980s, before people realized *a fed bear is a dead bear.*

As soon as I saw Thomas Mangelsen's iconic photograph *Catch of the Day*, grizzlies captured my interest. The picture—which has been called the most famous wildlife photo in history—shows a large brown bear standing on the lip of Brooks Falls in Alaska's Katmai National Park, and it's about to clamp its massive jaws on a sockeye salmon leaping up the falls. I immediately bought a poster of the photo, mounted it on posterboard, and hung it above my bed, next to pictures of basketball greats like Larry Bird and Michael Jordan.

It wasn't until after college that I finally encountered my first bear in the wild, in Washington State's North Cascade Range. My plan that afternoon had been to hike up to the old fire lookout atop Sourdough Mountain, where poet Gary Snyder spent a summer in 1953. On the way up, I ran into an off-duty park ranger named Dena. This chance encounter would change my life forever.

Dena was the first "bear person" I'd ever met. She was comfortable living and working in bear country and didn't mind incurring a little bit of risk because of the many benefits of having bruins on the landscape. She took pains to secure things like garbage, compost, bird feeders, and beehives around their house so as to keep bears wild and not attract them to human areas.

As Dena and I hiked down from the lookout that sunny afternoon, I rounded a corner and saw a black bear with tiny eyes, seated on his haunches, feasting on huckleberries.

"Bear!" I exclaimed, as the bruin stood tall.

"Don't run," said Dena, grabbing my arm.

"He's going to attack," I whispered nervously.

"He's just getting a better view and smelling us," Dena replied. "Stay calm."

She pointed out that the bear wasn't showing any signs of aggression or agitation. "He's just curious."

I was amazed. Dena seemed to speak the same language as the bruin, understanding and knowing how to interpret its behavior.

Moments later, as the bear grew disinterested, wandering away up the hillside, Dena led us down the trail.

"That was awesome!" I said. "My first wild bear!"

As we continued, descending over forty-five hundred feet down the steep trail with switchbacks, Dena told me about her job working for the National Park Service. She was an interpretative ranger who staffed visitor centers and gave campfire chats at Joshua Tree, North Cascades, and Isle Royale. She told me about her colleagues who manned entrance stations, handled maintenance, or served as law enforcement rangers and ambulance workers.

"Sounds like a dream job," I said. I was amazed at the idea that people could be paid to live and work in magical places like Zion, Glacier, Acadia, or the Great Smoky Mountains.

Consequently, years later, after I graduated from paramedic school, I eagerly applied to work at Yellowstone and was eventually hired as a summer seasonal paramedic in the Old Faithful District.

It was there that I encountered my first grizzly in the wild, crossing the Firehole River near Morning Glory hot spring.

A few tourists standing beside me happily snapped pictures as I freaked out and fumbled into my pack for my bear repellent and radio when I spotted the tawny brown fur, the shoulder hump, and the flat, dish-like face.

"Dispatch," I said, keying my radio.[2] "We've got an emergency at Morning Glory. There's a grizzly!"

The dispatcher copied my radio traffic. "How far away is the bear?"

"About a hundred yards!"

"Is it approaching people?"

"No, ma'am."

"Are people approaching the bear?"

---

2. Kevin Grange, *Wild Rescues*, (Chicago Review Press, 2021), 24.

"Negative."

"Is anyone hurt?"

"No."

I heard what sounded like a sigh over the radio, and I could literally picture the dispatcher rolling her eyes. "So, you called to tell us there is a grizzly bear in Yellowstone National Park?"

I felt like such an idiot. Even the tourists, city slickers from Boston, looked at me like I was crazy.

"Um, yeah," I replied. "I guess so."

"Are you in danger?"

I said no.

"Do you need additional resources?"

"Negative," I replied. "Sorry for wasting your time."

As I turned off my radio, the bear disappeared into the woods.

My radio transmission had been broadcast over the entire Old Faithful District. I felt utterly ashamed, but I was terrified by grizzlies.

Over the next five summer seasons I worked at Yellowstone and Grand Teton, I saw probably a dozen brown bears. But the bruins were often far away in the Lamar or Hayden Valleys or ambling along the distant tree line of Pacific or Pilgrim Creeks. I'd watch the bears through binoculars or from the safety of my car or the ambulance—as if at some drive-in wildlife movie theater—and then go home each night comforted by the thought that I lived near but not in bear country.

Things began to change in 2019.

One morning that October, I arrived at my home in the South Park area of Jackson Hole, bleary-eyed and exhausted after a forty-eight-hour shift at Jackson Hole Fire Department, to discover an orphaned, emaciated black bear cub perched in my cherry tree.

"We've named him Hissy because of the sound he makes," Taz, my ten-year-old neighbor, said. "He's been here for two days."

"He's been sitting in your tree during the days and sleeping under the deck," added Taz's mom, Heide. "We've called Game and Fish."

The wildlife team arrived an hour later, and Heide filmed the biologists as they darted Hissy with a tranquilizer then prepared to catch him in a large tarp when he fell from the tree.

When Heide later posted the video online, it immediately went viral, racking up millions of views on Facebook, Instagram, and TikTok. There was something so heartwarming and human-like about how Hissy dangled from the tree by his two forepaws with his legs hanging below as if participating in some kind of team-building exercise, seeming to say, "You guys will catch me, right? You promise?" before letting go and landing safely in the tarp.

Hissy spent the winter in a wildlife rehab facility in Idaho and was released into the wild later that spring as a healthy, 112-pound yearling. I'd assumed the encounter with Hissy would be a one-off incident, but little did I realize that he was a harbinger of things to come.

Two years later, in the fall of 2021, the grizzlies arrived.

It began with an email and a subject line reading: "SAGE MEADOWS HOA: Notice Regarding Bear 399 & Cubs."

Bear 399 was the scientific number biologists had given to a twenty-six-year-old sow born in Grand Teton National Park. She was also known as the Matriarch of the Tetons, famous for raising her cubs roadside. To the delight of millions around the world, she'd given birth to over eighteen cubs, across eight litters, around Pilgrim Creek since 2004.

Bear 399's fame grew in 2020 when she emerged from the den with four cubs, a rare feat in the ursine world. Sadly, celebrity also had a dark side: two of her cubs from previous litters were killed by vehicles over the years, numerous of her offspring were captured or killed due to conflict, and in the fall of 2020 the family of five got into unsecured compost, livestock feed, and a beehive south of Grand Teton and were fed molasses-enriched grain by an eccentric homeowner in the Solitude subdivision who believed she had a special bond with wild animals and routinely fed bears, moose, elk, and deer in her backyard. The problem with bears obtaining food from humans is they forgo natural foods and become increasingly emboldened. They gradually lose their fear of people and are more likely to attack someone or cause property damage.

Fast-forward to the fall of 2021. I opened the email from my home-owner's association and read, "399 and her cubs are not back north near Grand Teton National Park. The bear clan was spotted south of the town in a big area known as South Park. Please keep dogs on a leash, secure food and garbage cans inside home or garage. Be alert when walking outdoors and carry bear spray."

Apparently, the ursine clan was on an extended walkabout around Jackson Hole and Teton County. They'd strolled through the parking lot outside the police station, a block from the fire station where I worked. They'd wandered on the hillside above Albertson's grocery store where I shopped. They'd also been spotted on Josie's Ridge, my favorite hiking trail, and even at the rafting take-out spot on Snake River where Meaghan and I ended our river trips. I no longer relished the thought of living in bear country. Bear 399's presence felt personal, intrusive, and, to be honest, a bit predatory.

Turned out 399 and her four cubs weren't the only grizzlies near our home. Another friend spotted eight brown bears on my favorite mountain biking trail, Munger Mountain, and a big cabin-brown boar with white claws had raided my buddy's chicken coop down the street from us. The bear had to be caught in a culvert trap by wildlife officials.

After reading the email from my HOA, I leaped up from my computer and hurried downstairs. When I worked in Yellowstone and Grand Teton, I was terrified of grizzlies. Now, as they invaded my community, I considered them a dangerous nuisance.

"Asher, come inside!" I called to our golden retriever in the back-yard. "Let's go!"

It was overcast and windy outside, the way it looks in the movies before a big storm or aliens arrive. Once Asher was safely inside, I went into the backyard and grabbed our bird feeder.

"What's going on?" Meaghan asked as I tore through the kitchen. "What about my hummingbirds?"

"The grizzlies are here!" I exclaimed and raced back to the deck to bring inside two big white compost buckets, overflowing with eggshells, banana peels, coffee grounds, and wilting lettuce.

"This compost has to go!" I exclaimed.

"Huh?" Meaghan replied.

"The bears!" I declared. "We need to batten down the hatches!"

Meaghan watched in amusement as I ran between the back deck and garage, removing anything that a curious bear might find interesting. After my final trip to the garage, I gave her a can of bear spray. "Don't go anywhere without this!"

"You're insane," Meaghan said with a laugh.

"And we need to keep the dog on a leash at all times!"

Even Asher was looking up at me like I was crazy.

The next week felt like a Hollywood movie. Bear 399 and her four cubs toured most of Jackson Hole, getting into food and damaging property. Due to the onslaught of incidents and phone calls, Wyoming Game and Fish had to call in the feds. US Fish and Wildlife launched "Operation 399," which consisted of personnel "bear sitting" the ursine family 24-7 as they roamed around Teton County, trailed relentlessly by wildlife paparazzi photographers.

"We've had repeated conflicts over a three- or four-day period, way down south," Game and Fish large carnivore biologist Mike Boyce told the *Jackson Hole News & Guide*.[3] "Property damage, livestock feed and apiary damage. We're having grizzly bear conflict way down south. That's been a change."

The irony is that the town of Jackson Hole and Teton County is filled with people who love the outdoors, and we pride ourselves on protecting wildlife. Yet when it came to bears, we were not as up-to-date on bear safety as other communities like West Yellowstone, Montana, or Durango, Colorado.

"The whole county is kind of behind the times in terms of trash and storage and conflict prevention," said Hilary Cooley, grizzly bear

---

3. Mike Boyce, as quoted in Mike Koshmrl, "Operation 399: Feds Keep 24/7 Watch on Beloved, Embattled Bruin," *Jackson Hole News & Guide*, October 27, 2021.

recovery coordinator with US Fish and Wildlife.[4] "Beehives, livestock feed, open dumpsters. Almost everywhere you look there's something."

Fortunately, despite ten conflicts involving 399 and her cubs, the family eventually returned to Grand Teton National Park and went to den without having to be relocated or killed; however, six other grizzlies in the area—including three descendants of 399—had to be euthanized due to causing property damage and bypassing native foods in favor of raiding garbage cans, beehives, and chicken coops.

Am I a bear person? I wondered one day while gazing out at the Tetons. Now that they were in my backyard, I felt strangely conflicted.

I wasn't alone. After grizzlies were placed on the endangered species list in 1975, their recovery in the Lower 48 has been one of the great conservation success stories. As brown bears increased in numbers and expanded their range beyond protected areas, millions of people in the West found themselves at a similar crossroad. *What risks do grizzly bears pose? Is it worthwhile having them in the landscape? How can I stay safe in bear country? And, most importantly, how can we coexist?*

I wanted answers, but I also realized I needed more information. Beyond their size, strength, and perceived ferocity, I knew very little about brown bears.

A few weeks later, I was surfing the web when I happened upon a link for the Sixth International Human-Bear Conflict Workshop, held in Lake Tahoe in October, which promised four days of nonstop lectures, panels, and presentations about conserving bears and connecting people.

Seems like a good place to start, I thought, registering and booking a flight. I had no idea my relationship with brown bears and the natural world would change forever.

I thought I was just going to a conference.

---

4. Hilary Cooley, as quoted in Mike Koshmrl, "Operation 399: Feds Keep 24/7 Watch on Beloved, Embattled Bruin," *Jackson Hole News & Guide*, October 27, 2021.

Chapter One

The Leading, Bleeding Edge of Human-Bear Relations

Problem bears are not born, they are made.

—*Charles Jonkel*

I hurried through the cool air and carpeted room that smelled of cigarettes, past poker tables and slot machines with nature names like Buffalo Ascension, Wolf Run, and Panda Magic; hopped on an escalator down to the bottom of Harrah's Hotel and Casino in Lake Tahoe; and walked into the American Café to find the place packed and every table taken.

I couldn't believe the café at Harrah's—a fifteen-story hotel with 512 rooms and seven restaurants—was full at such an early hour on a Monday morning, but then I recalled the reason I was here.

Just then, a woman from a nearby table called to me. "Excuse me," she said, noticing me looking out of place. "Are you a bear person?"

It was a good question. Despite living in Jackson Hole and working as a paramedic in Yellowstone, Yosemite, and Grand Teton—all bear hotspots—I still wasn't sure if I was. Frankly, I didn't believe bears could be taught to avoid human space and be trusted. I was still terrified of grizzlies, doubted I would stand my ground if ever charged, and wasn't sure humans and bears could ever coexist, which was precisely why I'd traveled to Lake Tahoe in October for the Human-Bear Conflict Workshop.

"I'm here for the conference," I replied, dodging the deeper question.

The lady's face lit up. "We are too," she said, pointing to an empty seat at her table of six. "Have a seat!"

"Thank you," I replied, grabbing a chair at a table filled with a game warden, refuge manager, field biologist, and conservation officer.

The first Human-Bear Conflict Workshop was held in Yellowknife, the capital city in Canada's Northwest Territories, in 1987. The goal of the conference was to bring people from around the world who worked with the eight bear species to investigate conflict prevention and management. The workshop was held every few years at human-bear conflict spots like Canmore, Alberta (1997, 2009); Missoula, Montana

(2012); Gatlinburg, Tennessee (2018); and now, in 2022, at Lake Tahoe in Nevada. Before moving to Jackson Hole, Meaghan had lived in Tahoe and told crazy black bear stories, like when she once found a large sow in her shower, the time when a bruin broke into her car and made off with her toothpaste, and the bear that stole a vodka-infused watermelon from one of her parties.

"We're the ones who deal with most of the conflicts with bears, so it's up to us to find solutions," Linda Masterson, author of *Living with Bears* and conference co-chair, said as I picked up my name badge and swag bag at the registration table. "Make sure you register for the raffle to win a pneu-dart G2 X-Caliber," she said, showing me a photo of a satin-black projector (i.e., gun) with a thirty-nine-inch stainless steel barrel that, evidently, was perfect for remote drug delivery on large carnivores.

"Thanks, but I don't actively work with wildlife," I replied. "I'm writing a book."

After putting on my name badge, which outed me as a freelance writer, I wandered into the exhibit hall; it was full of vendors hawking the latest and greatest GPS collars, bear-resistant boxes, and portable electric fences. There was a scientific lab that offered tooth analysis; another business sold nonlethal ammunition, with names like "bangers" and "screamers," designed to scare bears; and another company sold a surveillance radar system to alert communities of approaching polar bears.

Just before a table full of coffee, muffins, and fruit, I was greeted by a black bear, mounted by a taxidermist. "You are here for me," the placard below stated. "I am a cautionary tale. I was getting into garbage. Residents were told by concerned citizens not to call agency biologists because 'they just kill all the bears.' I escalated my conflict behavior. Then I started entering cars until I entered my last one. The door closed behind me, and I was trapped. I died of heat stroke on a hot summer day. If the agency had the chance to capture me, they wouldn't have killed me. Instead, I died."

I grabbed a cup of black coffee, poured in enough creamer to turn it milky white, and headed into the large, windowless conference room filled with long, successive rows of tables.

The room was crowded with hundreds of people who carried laptops and notebooks in backpacks instead of briefcases. One woman drank from a water bottle bearing a bright orange sticker that said, "Keep Calm and Carry Bear Spray." Another man had a bunch of stickers covering his laptop computer: "Keep Bears Wild" and "Welcome to Bear Country, Please Be Bear Wise." And a woman to my right wore a red, Donald Trump–style hat with white lettering that read, "Make America Wild Again."

At eight o'clock Carl Lackey from the Nevada Department of Wildlife (and Linda's co-chair of the conference) took the stage to welcome us.

"Good morning and welcome," Lackey began. "It's good to be here, and it's nice to see so many people here despite the craziness we had to put up with during COVID for so many years. We're starting to get back together again, and that's really great."

Next, Masterson, carrying a blue Super Blast air horn, joined Lackey onstage where they informed us that since most conflicts were caused by people, we shouldn't use the terms *problem bear* or *conflict bear* for the duration of the conference.

"If you do, we're going to *honk* you," Lackey promised, as Masterson delivered a loud, 120-decibel blast from the air horn.

"This week will be filled with a diverse group of people—from bear managers to concerned citizens," Lackey promised. "So let's all learn to work together for the love of bears!"

Next, Lackey introduced Dr. Chris Servheen, who'd be giving a keynote lecture on the evolution of human-bear conflict in North America. Prior to retiring, Servheen worked for the US Fish and Wildlife Service for thirty-five years, serving as the agency's grizzly bear recovery coordinator. Servheen had initially been interested in bald eagles but became enamored with brown bears after watching a

National Geographic Society special in the 1970s. In the early 1980s, he served as a work-study student for the legendary professor and biologist John Craighead at the University of Montana in Missoula. Servheen, a bespectacled man in his sixties with a white handlebar mustache, took the stage and, during the next hour, distilled two hundred years of struggle into one riveting lecture.

The steps in the evolution were: kill the bears for being bears; kill the bears to save livestock; kill the bears to save deer, elk, and caribou; place livestock into bear habitat and kill the bears; intentionally feed bears at places like Yellowstone and Glacier National Parks; stop feeding bears garbage; kill the bears in national parks that are still getting into garbage; relocate bears that get into garbage; track bears to see if relocation works, and use deterrents like electricity and noise, bear dogs, and bear-resistant cans; begin to educate people about attractants; learn about bears and why they are there.

"So the journey has been from blame the bears to blame the humans," Servheen said. "From feeding the bears to keeping food away, from control and dominance over nature to attempts at coexistence, from valuing bears for amusement and trophies to valuing bears as animals representing ecosystem health, from respond to the problem after the fact to prevent the problems before they happen."

"Amen!" said a woman to my right who worked for Montana Fish, Wildlife & Parks.

"We've come a long way in how we think about and deal with bears and how we solve the problem of humans and bears. But we have a long way to go in our journey, though, as evidenced by a place like Lake Tahoe, where we have 150 black bears with sixty-five thousand people—and this hotel."

We all laughed because the dumpsters outside Harrah's Hotel and Casino were overflowing with trash, providing a nightly Thanksgiving dinner for local black bears.

Servheen continued: "This is what I call the leading, bleeding edge of human-bear relations, and these things don't end well. So that's why we're all here today, so we can better understand how we can success-fully live with bears and convince people to live with bears and change their behavior so the bears can exist successfully."

Following Servheen, John Hechtel spoke about the challenges of providing correct, consistent bear safety messaging. Hechtel had worked for the Alaska Department of Fish and Game for thirty years and was now president of the International Association of Bear Research and Management. Hechtel gave hilarious examples of some of the false claims over the years.

"For unexplained reasons, human couples engaged in sexual intercourse, especially during thunderstorms, have been attacked and killed by grizzlies" was one theory proposed after two fatal maulings in Glacier National Park on the same night in 1967. Another false theory: "Women have frequently been attacked because they were menstru-ating. This type of attack has usually occurred during the summer months when bears are breeding."

"Both claims are totally false," Hechtel said with a laugh.

From there, the workshop was a four-day, caffeinated blur of lec-tures presented by some of the world's foremost bear experts. I took copious notes as colorful and detailed graphs filled the presentation screens and terms like *X- and Y-axis, binary response, confidence inter-val, coefficient estimate,* and *sample size* were routinely tossed out. While I felt like more of an outsider—instead of the kind of *outlier* that would make Malcolm Gladwell proud—I learned a lot.

Shalu Mesari worked as a wildlife and conservation biologist in Gujarat, India, and in her presentation I discovered the world's most dangerous bruin wasn't the polar bear or the grizzly. "Sloth bears typi-cally kill over twelve people per year," she said, "many times by ripping their face off."

During Joe Savikataaq Jr.'s presentation, "Challenges of the Arctic," I learned how climate change was affecting people and *Ursus mariti-mus* around Hudson Bay. "With warming temperatures and less ice, polar bears, who are strictly carnivores, are less able to catch seals and other sea life and are being driven onto land, increasing the encounters with humans."

"Acknowledging the complex and often conflicting meanings given to the land, predators, and conflict-reduction tools may allow us to navigate a shared landscape and reduce conflict," said Allegra Sundstrom, a graduate student in sociology at Idaho State University. She was giving a talk titled "Reducing Livestock-Grizzly Conflict: Uncovering the Symbolic Meanings of Conflict," which explored how people think about grizzly bears, ranch-management practices, and the symbolic meaning on why some people adopt, or avoid, conflict-reduction tools.

"It's important to recognize conflict with grizzlies is situated within a larger landscape, and cultural shifts and decisions to adopt, or avoid, conflict-reduction tools often occur at these crossroads," she said. "Identifying trade-offs, whether symbolic or financial, that occur when individuals adopt conflict-reduction tools may be the first step toward finding the right tool for each ranching operation."

Heather Reich, from the Nevada Department of Wildlife, discussed the use of Karelian bear dogs to haze black and brown bears that were conditioned to getting food from humans, and a biologist from Minnesota researched the efficacy of hunting for controlling human-bear conflict.

"Hunting doesn't necessarily lower conflict, but it improves people's tolerance for bears and lowers complaints," he said, reading off a slide. "It also shifts unregulated killing to regulated and has been shown to improve people's acceptance for bears, even if they do occasionally cause damage or injuries. Hunting isn't always popular, but it has a role in human-bear conflict and provides economic benefit to local communities."

I discovered there were dozens of tips, tactics, techniques, and tools to reduce conflict: bear-resistant bins and dumpsters, electric fencing to secure chicken coops and beehives, picking fruit before it ripens and bears get into it, planting flowering trees instead of fruit trees, using bird baths instead of bird feeders (because they don't need food in the summer), utilizing Karelian bear dogs, instituting cattle-carcass pickup programs and compensation when livestock becomes deadstock, and using range riders, or human patrols on ranchlands that determine the presence of predators and identify any sick or injured cows or sheep, likely targets for a bear or wolf.

At the end of each lecture, there'd be five minutes of Q&A where occasionally someone would say the forbidden phrase and get honked at by the air horn.

"Hello, Mike Schultz with Fish and Wildlife," a middle-aged man in jeans and a flannel shirt began. "Has the use of electrified fences on farms improved the tolerance of ranchers to having grizzlies on the landscape and decreased conflicts with problem bears?"

"Honk him!" we'd all shout as Masterson blasted the air horn, to great applause and laughter.

Attendees were all there to help bears, but that didn't always mean they had similar means to achieve the same goal. Some people thought a bear that was habituated, or comfortable, around humans wasn't a bad thing because it took the edge off, likely reducing an attack if there was a surprise encounter with a human. Others thought habituated bears in public lands like a national park was bad because the bears were virtually guaranteed to wander onto private land at some point. "Problem is, particularly with males, bears disperse and end up leaving protected areas and start taking that behavior to the Wild West of private land, and they just don't last very long—they get food-conditioned and shot," a biologist said.

Some bear managers had a lot of success using aversive conditioning (AC), such as loud noise or nonlethal ammunition, to push bruins out of public spaces, while others felt an AC program took too much time, money, and manpower. "An aversive conditioning program needs to be in place 24-7 if it's going to be successful," one game warden declared. "You can't only push bears out during the day, because they will just become nocturnal."

Perhaps the greatest debate concerned roadside bears like Bear 399, the Matriarch of the Tetons. Should these bears be hazed away from the roadway corridor because it is easier to manage one bear instead of hundreds of people? Or should roadside bears be allowed to forage because they are ambassadors for the species and, technically, aren't doing anything wrong?

"It's not fair to haze a bear for just being a bear," a game warden said. "And I think managing people is easier than trying to change the behavior of a wild animal."

Despite the differences on certain issues, we always arrived at a better sense of the question, and some fundamental truths emerged: With the number of bears, and their range, increasing—along with the number of people living closer to wild places and recreating outdoors—the number of human-bear conflicts is bound to rise. Conflicts will also rise when it's a bad natural food year, such as a reduced salmon run or berry failure. For bear-resistant products to work, cities, towns, and campgrounds need to make the right thing easy, accessible, and effortless for the people to use them. People will never become tolerant of bears if they're busting into sheds, raiding chicken coops, killing livestock, and freaking out kids. Reporting and recordkeeping are key to determine the success or failure of conflict-reduction programs. Relocating bears is time-intensive and costly, and it rarely works because there's not some never-never land we can take bears to where they live happily ever after. The solutions to conflict will vary greatly depending on the location, type of bear, terrain, and community, and most bear biologists would rather crawl into an occupied den than speak with a journalist.

Suddenly, I felt all eyes on me and my name badge, proudly announcing me as a freelance writer.

"Part of the reason we get into biology is we don't like people," a bear management biologist joked.

"I can appreciate that," I replied, standing. "But this conference has shown the importance of public messaging and getting those positive stories out there."

Some of my favorite moments in the workshop occurred between presentations when we'd all gather in the exhibition hall to refill coffee, snack, and talk bears.

Nate Svoboda is a biologist with the Alaska Department of Fish and Game on Kodiak Island. He'd grown up in Nebraska and earned his BA in biology at the University of Nebraska before earning his master's at Central Michigan University. Following that, he worked as a wildlife biologist for the Little River Band of Ottawa Indians, studying black bears, wolves, coyotes, and bobcats, and eventually earned his PhD in conservation biology and accepted a job on Kodiak.

When I asked Svoboda what he found most fascinating about brown bears, he began with their resiliency and ability to adapt. "They have existed historically from the southern US and into Mexico, all the way north to northern Alaska and Canada, including parts of Russia and Europe, so that just speaks to their climate and ecological flexibility. Not a lot of animals can do that."

Svoboda thought a bear's sense of smell, many times that of a bloodhound, was cool, but he was most impressed with the concept of delayed implantation. "Bears typically breed in the spring, May and June, but they don't get pregnant per se," he began. "The egg that is fertilized in the breeding process develops into a blastocyst, but doesn't attach to the uterine wall until much later in the year."

"What determines that?" I asked.

It is based on their physical condition and, specifically, the maternal condition. "If the mom is in really good shape, then the blastocyst will attach to the uterine wall and pregnancy will take place as normal, but if she is nutritionally stressed or not in good shape, then she aborts that egg and implantation doesn't take place," Nate said. "That is a pretty unique thing that a lot of people don't realize either. They think they breed in the spring and gestation takes place for nine months."

"But you're saying it's much shorter?"

"The actual gestation, or growing, is only a couple of months, something like around sixty-four days. Bears are the only mammals in which delayed implantation, active gestation, birth, and lactation have been observed during hibernation."

February 1 is the unofficial birthday for grizzly bears in the wild, which typically give birth between January and February. The newborn cubs are all born premature, hairless, and helpless with fused eyes and ears, about the size and weight of a bean bag. The sow is hibernating when birth occurs, and the den acts as a kind of isolette and neonatal intensive care unit as she trades placental nourishment for her milk, high in fat, and nurses them into good health.

During another break, I met Joy Erlenbach, who'd earned her PhD at Washington State and was now a bear biologist at Kodiak National Wildlife Refuge. During her time at WSU's Bear Center, Erlenbach had spent around three hundred days living among the brown bears on the remote coast of Alaska's Katmai National Park.

"When I started working in Katmai, my last field job with brown bears had been working on a research project in Yellowstone, where you expect a bear to be afraid of you and run away, and if it doesn't, you are very concerned," Erlenbach said. "And it took me the first couple of weeks when guides were telling me I should kneel and sit and let the bears come to me. I told them they were crazy, and they said you're so wrong, this is the natural way bears should be behaving, and what you are used to is the unnatural way."

"What did you learn from the experience?" I asked, taking a bite of a blueberry muffin.

"It taught me that we can coexist peacefully with these animals, and they are not out to get you all the time," she replied.

At Hallo Bay, Erlenbach and her coworker slept in tents in a grassy area, and sometimes there'd be twenty bears grazing in the sedge meadows nearby. To remain safe, they chose a tent site out in the open and surrounded the area with electric wire, and they had multiple tools if they needed to confront a grizzly—bear spray, a flare gun, air horn, and gun. However, most of the time, simply talking, clapping, or shaking an empty garbage bag in the wind was enough to send a curious bear dashing off.

"Still, weren't you a little frightened?" I asked.

"You learn to speak the language pretty quick," Joy replied. "A bear's world is ruled by body language."

Some of Erlenbach's favorite moments in Alaska occurred when thousands of spawning salmon were running up the river and the bears had eaten their fill. "That's when they have time to play because they're not food stressed," she explained. "After they were full, they'd just be like floating in a lazy river, playing with salmon carcasses on their belly. I'd get to see all kinds of weird, goofy behaviors they do when they have spare time on their hands. That's where their personalities come out. They have huge personalities, and I really wish people got to see more of that."

Erlenbach certainly had a few close encounters living among a high density of bears, including rounding a bend and being so close to a sow, she had to *shoosh* the two cubs sleeping on their mom's back to not give her away while she quietly backed up. The encounter, and all the others, ended well. "I have run into bears and you both kind of go, 'Oh shit!' and then you sort it out. I just love the stories of bears walking the other way or walking the same way and just continuing along and walking past you. People just don't realize that's an option for a bear. They think if you run into a bear, it will attack you, and that's not the case. You don't hear about all the times people run into bears and it doesn't go badly."

Erlenbach added that there are always some bad apples and occasionally bears can wake up on the wrong side of the bed. "They could have just ten seconds ago been attacked by another bear and then take it out on you. I'm not delusional. They're still predators. They can still kill you, but I wish bear encounters gone well were told more."

Just then, the lights flickered, signaling the next session was about to start. Before she left, Erlenbach told me the best viewing is when you're down at the level of the grizzly in their natural habitat and you get to look them in the eye without infrastructure like a viewing platform, and the feeling of safety that comes with it.

"I really hope you get to experience that," she said. "If you don't get to experience coastal Katmai, Lake Clark, or McNeil River State Game Sanctuary, you're missing a huge, huge piece of the puzzle."

One of my favorite speakers was Jay Honeyman. He'd worked for ten years as a large carnivore conflict biologist with Alberta Fish and Wildlife. Before that, he'd served for seventeen years as a conservation officer with Alberta Parks and had also done some private contractor work with the Wind River Bear Institute, conducting grizzly bear conflict work with Karelian bear dogs.

Honeyman—whose last name is perfect for a bear biologist—spoke about the need for biologists and managers to look outside their own bubble and work on a landscape level instead of a jurisdictional level; he also discussed how ranchers carried most of the cost for human-bear conflict-reduction measures.

"There needs to be a discussion about the idea of having this cost shared among the greater public," he began. "The people who want bears on the landscape need to realize that private lands are a really important part of grizzly bear range, and if they want to have grizzly bears on private lands, they need to help out the ranching community with the costs that are borne by them to allow grizzly bears to be on their property."

Honeyman ended his presentation by speaking about the value of partnerships. "We couldn't do the work we do without community support. If you take the community piece out of it, we're dead in the water. We don't have enough resources to come close to what the community groups are doing. And education is a big part, and enforcement needs to also be a component. Most people get the education but there's always going to be that 10 percent that don't get it, and that's where enforcement kicks in."

The last featured speaker, Seth Wilson, was executive director of Blackfoot Challenge, a nonprofit that works to conserve and enhance the natural resources and rural way of life in Montana's Blackfoot watershed, a 1.5-million-acre swath of land that begins on the Continental Divide and ends at the Clark Fork River, east of Missoula, Montana.

Wilson spoke passionately about his lessons learned from working in a shared landscape and the "messy mosaic of land" that includes federal, state, and tribal agencies; private businesses; and landowners.

"We talk a lot about focal species in biology, but I think it's important to talk about focal people—key opinion leaders in the community who build credibility and trust. I think we put too much emphasis on fixing with the tools when it's about building relationships, communication, partnerships, and having some sort of decision-making apparatus that gets it all done. That builds trust in the community. We've also avoided taking positions on controversial issues, and that's kept communities coming back to the table."

Before he left, Wilson advised us to think of ourselves as expert learners and not *the* expert. "And if you carry that humility with you in the face of complexity, it will resonate with people and it will help build long-term relationships, and people will welcome you into the room because you're a great person to work with."

Wilson thanked us all for attending and then said, "Please invest in the long game. We need you."

As I left Harrah's Casino Thursday afternoon, I decided attending the conference had been a resounding success. Over the span of four days, I'd gotten a better sense of the history, causes, and current trends in conflict; learned about the new and proven methods of reducing problems; met a host of inspiring folks in the grizzly community; and been inspired to do my part to reduce problems with bears.

I was just about to hop into my rental car, parked outside Harrah's Casino, and drive to the Reno airport when I noticed trash spilling out of a half-open dumpster. Donning the pair of EMS exam gloves that I always carried with me in case of a medical emergency, I jumped out, collected the trash, and threw it back in the dumpster before closing the lid and shutting the bear-resistant latch.

Chapter Two

Keep Calm and Carry Bear Spray

If we can learn to live with bears, especially the grizzly, and if
we can learn to accommodate the needs of bears in their natural
environment, then maybe we can also find ways to use the finite
resources of our continent and still maintain some of the diversity
and natural beauty that were here when Columbus arrived.

—*Stephen Herrero*

Peter Mangolds

Todd Orr was mauled by a brown bear—twice in the same day—and lived to tell the story. The first attack occurred just after dawn on the north fork of Bear Creek in southwest Montana. Fifty-year-old Orr had risen early on October 1, 2016, drove from his home in Bozeman, and hiked three miles into the backcountry to scout elk for the upcoming hunting season. Winter was coming and he needed to fill his freezer.

Orr had grown up in Ennis, Montana, just thirty minutes down the road at the base of the Gravely Mountains. Since childhood, Orr had loved being outdoors. Growing up, he'd spent every weekend in the mountains. He'd caught his first trout on the world-class Madison River at age seven and shot his first elk on opening day when he was twelve. Later he'd gone on to study fish and wildlife management at Montana State University in Bozeman and assisted with research on bears, wolves, and mountain lions in Glacier National Park and on coyotes, wolverine, and elk in the Greater Yellowstone Ecosystem.

After school, Orr was hired by the US Forest Service and had spent the last thirty years specializing in surveying and constructing trails in the Custer Gallatin National Forest, which occupies over three million acres along Montana's southern border.

As Orr followed the meandering trail that morning, dawn turned the surrounding forest gray, and Sphinx Mountain was a looming shadow in the far distance. No stranger to bear encounters, Orr kept his head on a swivel as he hiked up the creek, yelling "Hey bear!" every few feet.

Despite the early hour, Orr felt confident and safe. A can of bear spray hung beneath the chest harness holding his binoculars, and a high-powered handgun was holstered on his hip. He'd also taken bear safety classes in the past and had practiced deploying both his bear spray and pistol, so the quick draw was unconscious and immediate.

Three miles in, Orr spotted a brown bear sow with two cubs on a ridge a hundred yards away. He froze. The adult female sniffed the brisk October air for a moment, then turned and disappeared over the ridge with her cubs. Orr waited a few minutes after the family of three was gone, then continued hiking in the opposite direction.

Suddenly, he heard branches breaking behind him. He turned and spotted the sow charging down the hill toward him. Low to the ground with her ears pinned back, she was in an attacking position.

Orr tore the bear spray from his chest holster and yelled loudly, hoping the charge was only a bluff, but the bear kept coming—a freight train of jaws, paws, and claws.

As the grizzly closed in, Orr deployed his bear spray, creating an orange cloud of capsaicin between him and the charging bruin. But the sow's momentum carried her through the spray, and she was on him, bearing down and biting.

Using his backpack as a shield, Orr rolled face down to protect his internal organs, and he clasped his hands over the back of his neck and drew his elbows close to his ears to protect his head. Wheezing from the bear spray, the grizzly unleashed a flurry of bites to Orr's back, shoulders, and arms. And then, the sow abruptly released him, disappearing again into the forest.

Once the bear was gone, Orr spit dirt from his mouth and struggled to catch his breath. He rolled over to assess his wounds. Luckily, the bear hadn't cut any major arteries or veins, but deep puncture wounds decorated his back, shoulders, and arms, soaking his clothes in blood.

Get to the truck, Orr told himself, grabbing his bear spray and stumbling to his feet.

Orr moved down the trail, hoping to put distance between himself and the bear. As the adrenaline subsided, his whole body ached as if he'd been stabbed repeatedly with an ice pick.

Again came the sound of something moving through the woods. Orr turned, but before he could grab the 10mm pistol from his hip, the grizzly was on him a second time. Orr heard the crunch of his bones breaking and tendons snapping as the sow slammed him to the ground and once again buried her canines deep into his back, shoulders, and forearm. He again stayed tucked in his defensive position, lying prone with his hands interlocked behind his neck, but the bear's claws caught his forehead, leaving a yawning, blood-soaked laceration above his right eye.

The pain made him cry out—and the sound of his distress made the sow bite harder, lifting him up and slamming him to the ground.

Don't move, Orr told himself, struggling to stay optimistic. *She's going to leave.*

After a moment, the bear paused her attack. Standing atop Orr's outstretched body with all four paws, her claws dug into his back. Orr could smell her pungent breath blasting him with warm, rancid exhales.

And then the bear was gone.

Orr slowly rolled over and assessed himself once more. The injuries were more severe this time: his left arm was a mangled mess of bone, tendon, and muscle, and blood oozed from the laceration on his scalp, stinging his eyes. Yet, incredibly, he had no life-threatening injuries.

Orr gathered his bear spray and handgun, both of which had been torn from his body during the melee, and started back down the trail again. Forty-five minutes later, Orr reached his truck. His first thought was to leave a "bear in the area" note at the trailhead to warn other hikers. But he was too amped up and shaking to write, so only blood and an unintelligible scrawl appeared on the page.

Get to the hospital, he told himself.

The closest hospital was Madison Valley Medical Center in Ennis, thirty minutes away. Orr knew he had to hurry since the golden hour—the ideal window after a traumatic injury to get surgery—had already passed. But not before he took a few photos and a video for his hunting buddies.

Orr brought his phone to his face and hit record. "Life sucks in bear country," he began wryly, his face streaked with blood. "Just had a grizzly with two cubs come at me from eighty yards. She got my head good. I don't know what's under my hat, my ear, my arm. Pieces of stuff hanging out, and I don't know what's going on in there. And my shoulder is ripped up. Think my arm's broke. Internal organs are good. Eyes are good. I just walked out three miles to the road and now I have to go to the hospital. Be safe out there." Orr stopped recording, then hopped in his truck and started for the hospital.

He arrived at Madison Valley Medical Center half an hour later to find medical staff and a local sheriff waiting for him. On the drive over, Orr had wisely pulled over and asked a rancher to call the hospital and alert the trauma team that a bear attack survivor was en route.

When Orr slowly pulled into the hospital, a police officer helped him place his truck in park, remove his seat belt, open the door, and stumble into the hospital. Two doctors spent the next seven hours cleaning Orr's wounds and stitching him back up. The following morning, he underwent surgery to reattach the shredded muscles and tendons of his left arm and had major reconstructive surgery done on his shoulder. Orr was told that while rehabilitation would take months, he was lucky to be alive.

Several years after his attack, I spoke with Todd Orr about whether his relationship with the outdoors or grizzlies had changed.

"My life has always been in the outdoors, so I knew I had to get back out there as soon as possible," he told me. "The day my shoulder brace was removed, I ventured back to the mountains for the final four days of hunting season, hoping to put an elk in the freezer."

"Did you see any grizzlies?" I asked.

"The first day back, I encountered fresh bear tracks within the first hour. I was certainly not comfortable with the situation, and my head was on a swivel. But I pushed through the uneasiness for a ten-mile hunt. The following day, with a buddy in tow, I went back to the location of my attack. Another uncomfortable day, but I knew I had to push through it as soon as possible."

The following spring, Orr went back into the woods for work and ran into a sow grizzly with a young cub at fifty yards, but fortunately she ran in the opposite direction.

Over the following years, Orr has become comfortable working in bear country again but says he's always paying close attention to his surroundings in hopes of avoiding any future up-close encounters. "To this day, I still enjoy seeing bears across the landscape, but just prefer them at a distance," Orr said. He emphasized that people venturing into the area should also pay attention to their surroundings, not wear earbuds, make noise while hiking, and travel with a partner or group. "I believe that anyone venturing into the outdoors should have some knowledge of bear identification, awareness, behavior, and safety. Even better if you can attend an instructional class and learn how to properly deploy and use bear spray."

Following the attack, Orr focused on his side business as a custom knife maker. Skyblade Knives is a Montana-based company focusing on handcrafted knives and cutlery—"Art with an edge."

"I do support a management plan to limit the growth and population of grizzly bears. Bears roam throughout much of this country, with human-bear encounters inevitable," Orr said at the end of our conversation. "Ninety-nine percent of the time, you don't have an issue. The bears don't want to attack people. They don't want to eat people. They're only a problem when you're a threat to them."

As for his own survival story, he said, "It's quite amazing what the human body and mind can do in a survival situation."

The photos and video Orr recorded of his attack and posted online once he left the hospital immediately went viral on social media, and all the major news outlets wanted an interview with him. There was something about Orr's courage, calm resilience, and survival skills that inspired millions, and he was called a hero. Orr used the incident as a platform to teach people about staying safe in bear country, citing the pioneering work of Dr. Stephen Herrero.

Peter Mangolds

The bible on bruin encounters is a book titled *Bear Attacks: Their Causes and Avoidance* by Dr. Stephen Herrero, a Canadian emeritus professor of ecology at the University of Calgary.

As with John and Frank Craighead, who employed the scientific method to study grizzly bears in Yellowstone in the 1960s, Herrero used a similar tactic to shine a light on what prompted bear maulings, killings, and close encounters of the ursid kind. He asked, "Could the methods of science be used to reveal the causes of bear attacks?"[1]

"My aim isn't to horrify but to describe attacks significantly to characterize and explain them," he wrote. "And then suggest how to minimize them."[2]

Like a crime scene investigator for carnivores, Herrero asked eight important questions: What happened? What was the time, date, and duration of the incident? Where did it occur? How did it happen? Did anyone witness it? Was it a bear attack or simply an encounter? What did the victim do in response to the incident? How did the bear behave?

---

1. Stephen Herrero, *Bear Attacks: Their Causes and Avoidance*, (3rd ed. Essex, CT: Lyons Press, 2018), XVI.
2. Herrero, XVIII.

Some key findings quickly emerged. First, Herrero pointed out that millions of people ventured into bear country each year for hiking, camping, and other leisure activities without incident.

Herrero also theorized that black and grizzly bears may have evolved differently to respond to perceived threats. Black bears, native to forests, can climb trees or disappear into the woods when threatened. Brown bears, native to open areas—arctic tundra, meadows, the plains, and ice sheets—often lack the shelter of forests so respond to danger by charging forward and attacking.

According to Herrero, the best defense is a good offense. "If you see a grizzly before it sees you, and you can retreat without attracting the bear's attention, then you are acting as safely as possible."[3]

Herrero looked at twenty years of detailed records of human-bruin encounters—both black bear and grizzly—of which 357 out of 414 involved injury or aggression.[4] He broke attacks down into two categories: defensive and predatory. According to Herrero's theory, Todd Orr suffered a defensive bear attack that fateful October morning in 2016.

In a defensive attack, a grizzly perceives a human as a threat. This generally occurs in three situations: a surprise encounter with a bear at close range, encountering a sow with cubs, or coming upon a bear on a food source such as an elk carcass.

Todd Orr's incident involved a mother with cubs. When John Wallace, a fifty-nine-year-old man from Michigan, was fatally mauled by a grizzly bear on Yellowstone's Mary Mountain trail in 2012, a bear protecting a bison carcass nearby was to blame. When Craig Clouatre, a forty-year-old outdoorsman, husband, and father of four, was killed while searching for elk antler sheds on March 23, 2022, in Montana's Absaroka Mountains, north of Yellowstone National Park, it was a surprise encounter at close range.

In a defensive attack, the bear senses danger. The best course of action in that scenario is to show you're not a threat by slowly retreating

---

3. Herrero, 77.
4. Herrero, XX.

while talking in a low, calm voice. If the grizzly charges to within thirty feet, deploy bear spray. And if the grizzly makes contact, follow Orr's methods of "playing dead" by lying on your stomach to protect your internal organs, using your backpack as a shield, clasping your hands on the back of your neck with elbows close to your ears to protect the sides of your face, and avoiding making any noise that might mimic prey.

The other type of attack is a predatory attack. In these rare cases, the bruin perceives the human as food. Often, the bears responsible for predatory attacks are old, injured, or sick and unable to obtain nourishment by natural means. The best defense in a predatory scenario is to fight back in any way possible with loud noises, bear spray, or firearms. "Under such circumstances, playing dead would be akin to offering yourself to the bear," Herrero wrote.[5]

When a bear rampaged through the Soda Butte Campground in Cooke City, Montana, in the middle of the night in 2010, killing one man and injuring two others, it was a predatory attack. In July 2021, when Leah Davis Lokan was dragged out of her tent in Ovando, Montana, and killed, that was also a predatory attack.

The most dangerous bears are ones that associate people with food and have no fear of humans. These bears have "lost their natural wariness of people but retained their aggressive nature—a dangerous combination," Herrero warned.[6]

After speaking with Orr and hearing his story, I decided that if I was going to continue to live and recreate in bear country, I'd do everything in my power to minimize the risk of an encounter or attack.

Despite spending six seasons as a paramedic/park ranger at bear hotspots like Yellowstone, Grand Teton, and Yosemite, I'd only managed bear jams on the road by directing traffic, helped enforce a

---

5. Herrero, 68.
6. Herrero, XXI.

hundred-yard barrier between tourists and a yearling (which had taken refuge in a tree beside Old Faithful geyser), and been part of a group to haze, or push, a sow with cubs off the asphalt path leading to Castle Geyser. I hadn't had any close encounters with bears or been taught how to best defend myself at close range.

Which is how I found myself, a few weeks later, standing fifteen yards away from an angry grizzly bear about to hurtle in my direction at thirty-five miles per hour, with less than three seconds to deploy the bear spray attached to my chest harness in order to deter the threat.

Thankfully, it wasn't a real bear, but a simulated bear attack I was encountering during a one-day, nonlethal bear defense class Orr had recommended called "Surviving the Griz."

The class was taught by Chris Forrest, owner of TACTIC, a training center near Bozeman that specializes in empowering its students to learn "a new mindset that comes from being prepared, seeing options, and understanding human worth for the purpose of protecting yourself and others."

Forrest, an athletic, no-bullshit kind of guy in his fifties, is a former Navy SEAL, an expert on personal defense, close-quarters combat, and counterassault. He deployed three times to the Persian Gulf as a member of SEAL Team One and five times to Afghanistan as a protective security specialist. Since 2009 Forrest has taught classes at TACTIC on handguns, rifles, self-rescue, leadership, self-defense, first aid—and a highly popular class about surprise encounters with *Ursus arctos horribilis.*

What could a Navy SEAL teach about bear defense? A lot, I believed. Navy SEALs are experts in stress inoculation and performance under pressure. At TACTIC, Forrest trained his students in stressful situations, a process he called *forging.* While I could've attended a more casual bear spray demonstration like those held in city parks by US Fish and Wildlife with a fun carnival feel, I wanted to experience something as close to the real thing as possible. I wanted to be stressed and scared so I could learn how I'd respond. Along with writing books and outdoor-oriented articles for magazines like *National Parks*, I'd worked full-time as a firefighter paramedic with Jackson Hole Fire/EMS since 2015, where I learned the value of "pressurized training" and putting myself in uncomfortable training situations to better learn how to solve problems and execute under stress. I could storm into a burning building to put out a fire, run a call for a patient in cardiac arrest, and insert a breathing tube into an unconscious patient's windpipe, but I wondered, *Could I handle a bear encounter at close range?*

"Most people carry bear spray or a bear gun into the woods but have very little confidence with these weapons because they haven't trained enough," said Forrest at the beginning of class. "Training develops confidence."

The class was held one stormy afternoon at TACTIC's headquarters in Four Corners, Montana. Shaking off the rain, I entered a small warehouse space divided into a series of small, plywood rooms with an elevated walkway for the instructors. This area was also where TACTIC held their close-quarters combat (CQC) classes for law enforcement and interested civilians to learn how to kick down doors and clear rooms as part of a SWAT or SEAL team.

Our class began with a video of Todd Orr recounting his attack and the bloody video he'd filmed just after his mauling. Next, we watched a video shot by a passenger inside a car traveling thirty-five miles per hour with a grizzly bear running alongside it, keeping pace down the highway.

"After watching that video, does anyone still think they can outrun a brown bear?" Forrest asked rhetorically.

After discussing bear identification, especially the shoulder hump, and dish-shaped face of grizzlies and long claws used for digging, we watched more videos that showed the body language of stress in a grizzly—a stomping gait and vocalizations like woofing, huffing, and jaw popping.

"If the bear is clacking its teeth, sticking out its lips, woofing, or slapping the ground with its paws, you are way too close and should back away," Forrest advised. He spoke with the authority of a seasoned biologist, part of the Navy SEAL principle of taking extreme ownership of information, skills, and tactics. "If the bear's ears are forward, it's likely curious or trying to show dominance. If the ears are flattened back, it means the bear is feeling stressed and unhappy."

We were handed grizzly bear and black bear skulls to examine as we compared and contrasted the two species based on their physical attributes, behavioral characteristics, mental acuity, sources of conflict, and the pros and cons of the various deterrents—bear spray, marine flares, air horn, and handgun.

"For all of these, context matters," Forrest explained. "And always think, two is one and one is none." This was the military concept of redundancy: we should always carry at least two kinds of deterrents in case one failed, broke, or was ripped away during a scuffle with an angry bruin.

"Other deterrents include traveling in groups, avoiding hiking at dawn and dusk when bears are most active, and always making noise so as not to surprise the bears," Forrest added.

Todd Orr's attack was discussed as a case study.

"What went well? What could he improve upon?" Forrest inquired. "And what are the lessons learned?"

"Despite suffering injuries, Orr did everything correctly," I offered. "He'd taken bear safety classes and properly identified it as a defensive attack."

A fellow classmate, a hunter whose favorite color appeared to be camouflage, raised his hand. "He carried bear spray that was easily accessible, and he'd practiced deploying it and also had a gun as a backup."

A hiker raised her hand next. "When Orr noticed the sow and cubs in the distance, he attempted to retreat silently in the other direction. When the bruin charged, he deployed bear spray. And when the sow made contact, he played dead and protected his vital organs, face, and neck."

"The only thing I might have done differently was to have the firearm in hand as he hiked down," said a fly fisherman in his fifties.

"But what are the odds of getting attacked twice?" I countered.

"It was just a bad day in bear country," said a thirty-year-old mountain biker.

Moving into the practical portion of class, Forrest discharged bear spray into a garbage bag, then cracked it open for us all to briefly inhale a whiff to become more comfortable with how bear spray works.

"Feels terrible," I managed, coughing. "Like someone's lighting your eyes and skin on fire!"

By then the rain had stopped, and Forrest asked us to go outside and watch as a student discharged a full can of inert bear spray. "What do you think about the reach and duration," he asked the class, "and how long it was able to deploy?"

"Didn't go nearly as far as I thought, so we're going to have to wait until the bear's a lot closer to deploy it," said a woman in her sixties who lived on a ranch.

"It only lasted a few seconds, and it was affected by the wind," added a hiker.

Forrest nodded. "Exactly. See why you should have a backup?"

We also learned how to spray: Not too high, because bears are often low to the ground when charging. Continuous pressure is better than short, quick bursts since creating an area of high pressure is what gives the deterrent its reach. It is best to deploy the capsaicin (composed of chili pepper oil) in a circular motion , or a Z-pattern, to create a protective cloud—not side to side, like Sylvester Stallone spraying bullets with his M60 in *Rambo: First Blood, Part II*. Lastly, bear spray is not to be applied to oneself like bug spray, which, some tourists have found out the hard way, would be not only painful but ineffective.

Next, we practiced our bear spray deployments. Thanks to his years in the military and experience running TACTIC, Forrest had a gift for creating high-stress scenarios and making his students adapt and overcome during unforeseen situations.

I was the first student to attempt the simulated attack. I felt all eyes on me as I assumed my spot, twenty-five yards away from a lifelike prop, consisting of a poster-size print of a ferocious, snarling bear. The poster was attached to a small, aluminum sled with wheels on a track and attached to a spring-loaded cable. My can of inert pepper spray was attached below the chest harness I typically wore when I hiked, which also held my camera and binoculars.

"Let's do this," I said, readying my hand near my holster and staring down the bear like some quick-draw gunfighter from the Wild West.

"Did you say you were from California?" Forrest asked casually.

"Grew up in New Hampshire," I replied, glancing over at him and relaxing. "I lived in California for a while, and I'm now in Jackson Hole where I work as—"

Suddenly, there was an explosive sound, and out of the corner of my eye, I spotted the bear charging toward me.

Like his SEAL instructors at the Naval Special Warfare Center in Coronado, California, Forrest had succeeded in distracting me, making me lose situational awareness. I was shocked and surprised to suddenly find myself in the midst of a grizzly attack.

Fumbling, I tore the bear spray out of my chest harness and flicked off the trigger guard, but I managed to discharge only a short burst before the grizzly landed at my feet.

"Dang," I said, sweating and peering down at the bruin. "That wouldn't have ended well."

The class chuckled. I felt embarrassed but also knew this was exactly the reason I'd signed up for the class.

"This is why we train," said Forrest, addressing the group. "Now, how could he have been more effective?"

The hunter said I'd done well by not having the bear spray tucked away in the back of my pack or under my coat. "Now the most important thing is to determine where you could draw fastest from—left hip, right hip, or a chest harness," he suggested.

The hiker added, "He could've taken off the trigger guard the moment he spotted the grizzly and saved a step."

"How about the orientation of the bear spray in the holster?" Forrest began. "In a real-world situation, he may not have time to draw the bear spray, so maybe if the canister is facing forward, he could deploy it from the holster and stop the attack?"

Forrest also told me I'd taken a step forward as I deployed the spray. "In this case, you don't want to close the distance between yourself and the threat," he said. "I'd suggest stepping to one side instead."

I was grateful for each new time-saving suggestion. As the grizzly prop was reset and another student assumed their place in front of the group, I vowed to keep practicing at home. Over the next few weeks, I worked on my bear spray deployment until I'd achieved quick draw automaticity through repetition. Between speaking with Todd Orr, reading Stephen Herrero's book, and training with TACTIC, I felt better prepared for my next hike in the Tetons or trip to Alaska.

Keep calm and carry bear spray.

Chapter Three

Charismatic Megafauna

You cannot share your life with a dog, as I had done in
Bournemouth, or a cat, and not know perfectly well that
animals have personalities and minds and feelings.

—*Jane Goodall*

Peter Mangolds

The enormous brown bear ambled toward me with her ears back, huffing and stomping her feet while giving a low, menacing growl. All signs of an imminent attack.

I should've been terrified standing five feet away from a beast with razor-sharp teeth and curved four-inch claws, glaring down at me from the top of the food chain. Instead, I watched in rapt amazement as a man called to her.

"Forward," the man shouted, egging on the eight-hundred-pound bruin with frosted silvery brown fur. "Good g-i-r-r-r-l!"

The brown bear paused at the knee-high safety perimeter separating us. Standing well over six feet on two legs, she could have easily jumped or casually stepped over the charged wires, but I felt absolutely no fear.

The man cast a quick glance at me. "See how I'm not requiring her? I'm *asking*," he said, before gazing back at the bruin. "Now b-a-a-c-k."

With that, the bear retreated, her wild eyes locked on us.

In the past, this leading lady, a celebrity sow named Honey Bump, had shared the screen with Brad Pitt in *Growing Up Grizzly* and Steve Carell in *Evan Almighty*, but today she was playacting with her owner, Doug Seus, the world's most famous brown bear trainer.

Doug was in his early eighties and had a frontiersman vibe about him with his gray grizzled beard and a red-and-black scotch flannel shirt tucked into faded blue jeans. He spoke in a booming baritone and was equal parts bear whisperer and shaman. His lesson with Honey Bump that afternoon—part instruction, part incantation—was like watching a magical dance.

"Good g-i-r-r-r-l!" Doug exclaimed, tossing Honey Bump a meatball, which she caught in her mighty jaws. "You're so g-o-o-o-d!"

As he threw another meatball, he explained, "The bears are like children. They need parameters. They have to mind their manners and not take advantage of a situation, but I'm always impressed by their intelligence, cognitive ability, and abstract thought."

We stood in a beautiful landscape of rolling hills, lush trees, and perfectly manicured grass, all framed by an unobtrusive fence to keep the bears inside. When not doing enrichment or training in the yard, the grizzlies here resided in large, clean pens, covered on four sides by a chain-link fence.

"The bottom line is I just love the beauty, the magic, the majesty, everything about bears," Doug said. "Grizzly bears are just the epitome of wilderness with a good-ass soul. Every adjective you want to use, they have it!"

After Honey Bump demonstrated an angry bear on the verge of imminent attack, Doug added additional behaviors. When she heard the command, "wide," Honey Bump opened her expansive jaws; she stood tall when Doug said, "up," and "circle" sent her rolling down the hill.

My favorite command was "yoga," whereupon Honey Bump balanced on her haunches and held the outside edges of her feet with her upper paws.

"It's happy baby pose for bears!" I remarked.

Doug and Honey Bump wanted to show me another move.

"Happy bear!" Doug bellowed, jumping up and down.

The giant grizzly followed suit, bursting into movement, leaping, bouncing on all fours, and running in circles like a puppy with a frantic case of the "zoomies."

I told Doug I hadn't realized that his work would be so physical.

"It is physical," he replied, his eyes locked with the grizzly's wild gaze. "And emotional. We are on a primal basis with one another. It's a totally honest encounter. There are no pretensions."

I'd driven four hours from Jackson Hole, Wyoming, to Heber City, Utah, that afternoon to spend a day with Doug and his wife, Lynne, legendary

Lynne Seus

trainers and conservationists who'd spent the last forty years living and working alongside grizzlies.

An hour prior to Honey Bump's comedic case of the zoomies, I'd pulled down a quiet residential street and parked in front of the Seuses' home, which looked magically transported from New England to the Beehive State. A white picket fence and porch framed a gray colonial house with a ribbon of black smoke curling from a brick chimney and a yard bursting with flowers of every shape, size, and color.

I followed a stone walkway over a burbling creek and arrived at the front door, knocking twice.

Moments later, Doug appeared. "Welcome," he said, giving me a big hug. "Good to meet you!"

"Come in," added Lynne. "You must be famished after the long drive from Jackson. Can I get you coffee or something to eat?"

If Doug exuded a frontiersman's vibe, Lynne reminded me of *The Brady Bunch*'s Carol Brady. She was kind and doting, wearing blue jeans and a wool sweater, adorned with a colorful stone necklace, and she spoke with a sweet, high-pitched voice.

I'd only just met the Seuses, but I was already impressed by their warmth, kindness, and generosity. Were they always this friendly with strangers? Or had they anticipated that I was a bear person, which meant I was now family?

"Coffee, please," I replied, marveling at their beautiful home full of antique furniture and wood floors that made wonderful, musical creaks as we walked to the living room, where the walls were decorated with wildlife art, Native American tapestries, flint arrowheads, and knives with deer-antler handles and obsidian blades. It wasn't immediately evident this was the cozy den of two esteemed bear whisperers until

I spotted their family photos above the fireplace, showing multiple generations: Doug and Lynne, their children, their grandchildren, and—upon closer inspection—grizzly bears, photobombing in every background.

"That's a picture of Big Bart as a cub lying next to our baby Sausha," Lynne noted, pointing to an image of a bear cub lying next to a newborn in a car seat. "They both weighed six pounds at the time."

As the family cat plopped down on my lap, Doug and Lynn spoke about how they got their start training wildlife.

"We met an old government trapper whose job was to trap vermin from the west desert," Lynne began. "But instead of drowning them, he'd bring them to us. We had just about every little critter you could think of."

The Seuses obtained a license from Utah Division of Wildlife Resources, and soon their home was a menagerie of not only dogs and cats but wolves, cougars, coyotes, deer, bobcats, and skunks, who occasionally got into the heating unit, infusing the house with a warm, noxious scent.

The Seuses enjoyed working with the animals, but their lives changed when they met Dan Haggerty, the actor from the 1974 movie *The Life and Times of Grizzly Adams.* "He was a really great guy, and we were in awe of him," Lynne said, her blue eyes twinkling. "I thought, if we're ever going to follow this dream, we need a Kodiak bear."

A few years later, in 1977, the Baltimore Zoo called. They had a newborn Kodiak cub that was "surplus" and asked if the Seuses wanted him. Despite having a newborn daughter, Sausha, they immediately agreed, and Bart the Bear joined their family. "He weighed only six pounds, and one eye was starting to unseal and his nose hadn't unsealed," recalled Doug, "so the first thing he ever saw, smelt, or heard was us."

Doug began training the bear immediately, and two and half years later, Bart landed his first role in the 1980 film *Windwalker* about a dying Cheyenne warrior telling his life story to his grandchildren.

By age five, Bart weighed over fifteen hundred pounds and stood over nine feet tall. The Seuses started Wasatch Rocky Mountain Wildlife to focus entirely on gently and humanely training brown bears for films, television, and documentaries.

"What drew you to grizzlies?" I asked.

"I was drawn to their immense capacity for deep, profound thought," Doug began. He spoke in starts and stops, slowly gathering his ideas, then reflecting and examining them. Listening to him speak, I was reminded of a guitarist hearing the first chords of a new riff or of a shaman summoning advice from the spirits.

"Viscerally and emotionally . . . bears are these big, powerful animals with cognitive thought . . . calculating . . . as smart as an ape . . . and as diverse in personality as any individual with mood swings," he said, adding, "There's a power there . . . it brings up your senses . . . you become feral. I mean, you have to, it awakens a sense of awe. They're just gorgeous, powerful, and beautiful!"

Lynne noted how philosophers believed the bear was our first deity. "Right at the crux of Neanderthal and Cro-Magnon, the caves in the back always had a bear skull," she said, remarking that the animal's long bones were often found placed through the eye socket, as if to ward off the spirit of the bear because the bear was so powerful.

"Every Native American nation I know refers to the bear as 'grandfather,'" added Doug. "If a bear is skinned, they look exactly like us—they have shoulders, shoulder blades, their knees go like this, their feet go like that. The only thing that's different is the claws."

We discussed pagan holidays that celebrated when the bears went to, and emerged from, their dens, and how bruins were depicted on many medieval shields. Lynne mentioned Eastern European legends of bears coming into the villages and capturing young maidens. "They'd take them back to their lair, and in the spring the woman would come back with this baby who'd become the fiercest warrior."

"Ever heard of the word *berserk*?" Doug asked.

"When someone is out of control," I replied. "How does it relate to grizzlies?"

"The word *ber* comes from the word *bear*," Lynne began, "and *zerk* means 'skin,' so when Viking and Swedish warriors went into battle, they'd wear bear skins because it made them *berserk*, giving them the strength and ferocity of the bear."

Conversation turned back to how they'd trained Bart and other bears over the course of their careers.

"Was there any specific technique you used or skill?" I asked, sipping my coffee.

"I've probably been blessed with a fairly good sense of timing," Doug replied. "It's a feel . . . a sensitivity to the animal. They're big and dangerous and powerful. They are what they are, and that's the beauty of them."

Rather than training bears, Doug described his methods as teaching them, but like any artist, it was hard to put into words just exactly what he did and how.

"They want routine and immediate gratification," he said of the bears. But withholding that was key. "So you break routine and make the break in routine positive."

To assist with this, Doug would always feed the bears at different times throughout the day, at different locations, and often using different-colored buckets.

"Patience is in them," he continued. "It's a process of how you teach them to evolve to patience."

"And then you reward them with a treat?"

"The reinforcement, the paycheck, is *emotional*, not food," Doug said. "It's affection."

I'd seen a video of Doug placing his face in Bart's mouth. "And you never think about getting hurt or killed?"

Doug said he was always in control. "It's absolutely necessary with a big predator to hold sway but hold sway with honesty and give them dignity."

Dignity was a big part of Doug's teaching process. Instead of giving commands, you ask the bear, because that, in his opinion, gives them dignity.

I was reminded of a phrase I'd heard. "Respect the bear and the bear will respect you."

The Seuses agreed.

Doug said there was an internal trust he had. "Don't be aggressive or demanding, but you have to be in control. You better hold sway, and that's the sense of timing. Everything that happens is in different degrees, and you must know what degree is infringement and what degree is demanding."

Trust. I hadn't thought it was possible between a human and an animal, especially a grizzly, but I was beginning to reconsider that idea.

"Grizzlies are as diverse of personalities as humans. You have super curious, mildly curious, and uncurious," said Lynne. "Profound, deep thought ones, and ones that don't think that heavily but still, in their own process, are extremely intelligent."

The hard part, she added, is being smarter than the bears, trying to keep up with what they're thinking and putting together and staying one step ahead. "We're talking about a wild animal that can track you, hunt you down, kill you, and eat you. That's at the top of the food chain; it's not us with our 7mm rifles."

Doug told me that all bears were curious by nature. "But the process of investigation is in the personality. Bart would rush in. Bart 2 was slower. All of them have enormous curiosity, but how they handle, react, or behave with that curiosity is different. And it can be different by the day. That's why you can't be too judgmental."

"You need to look at the status of their day or the ten days that happened before it," agreed Lynne.

Doug said Bart had seemed to know when he was on camera. "He'd get that look and lean real low, and his chest would be a little bigger and he'd look around and get that magic going."

Following his debut in *Windwalker*, other roles quickly followed for the bear. At home, Bart was fed wholesale chicken, fish, bear-essential daily vitamins, and produce donated from local grocery stores. "You wouldn't believe what America throws away," said Lynne. "A perfectly good apple is tossed just because it has one little spot."

When they traveled, Bart was bribed into good behavior like many children: with milkshakes, burgers, and rest stop swims in their kidding pool. "He became like a dog who enjoys going in the car and looking out the windows," Lynne explained.

The Seuses traveled to film sets across North America, Europe, and the Southern Hemisphere, hauling an air-conditioned trailer in back for their bears. They would stop for the night outside cities and towns, picking hotels next to big fields, perfect for a huge Kodiak bear to graze behind a safety permeter. If there was no vacancy at their favorite hotels, Doug and Lynne would enlist the help of locals who often crowded around to admire Bart at rest stops and gas stations.

"Where'd you catch him?" someone would ask.

"He's a film bear," Doug would reply. "Know any place where we could set up a fence?"

Dozens of hands would shoot up.

"My house!" one person would say.

"I can give you free coffee!" another would offer.

"Free coffee and breakfast," a third would say, topping the first two.

As Bart appeared in dozens of films, TV movies, and documentaries, he and the Seuses' reputation in the film industry quickly grew.

"Of all the movie stars I've ever worked with, Bart the Bear is as talented, cooperative, and charismatic as the best of them," said Ed Zwick, who directed the bruin in the movie *Legends of the Fall.*

Lynne said their bears have been taught by the world's best trainer. "They don't really do tricks. They have learned subtleties of behavior and character. They can show emotion and act emotion, and that's not a trick. They are actors in the finest sense of the word."

In the 1997 movie *The Edge*, directed by Lee Tamahori from a screenplay by David Mamet, Anthony Hopkins plays Charlie Morse, a billionaire whose plane crashes in the Alaskan bush with his rival, Robert Green, played by Alec Baldwin. Green is a business partner who fancies Morse's wife (played by Elle Macpherson), but the two have to set aside their differences to survive and overcome a bloodthirsty bear, played by Bart.

"Two consummate professionals who hold nothing back," said Kenneth Turan, an *LA Times* film critic. "[Bart's] performance here, the capstone of an illustrious career, is a milestone in ursine acting."

"Outstanding performance from Hopkins and Baldwin and one hell of a terrifying bear," gushed Lisa Schwarzbaum in *Entertainment Weekly.*

Bart could play comedy or drama and, like some Method-acting ursine version of Marlon Brando, he fully inhabited his roles, giving nuanced performances of surprising emotional depth—by turns vulnerable and volcanic—that rewrote the script on what animal actors could achieve.

"I'll never forget that film with Redford and Morgan," Doug said, meaning Robert Redford and Morgan Freeman.

"The movie *An Unfinished Life*," clarified Lynne. "That last scene, with the encounter between Morgan Freeman and Bart, was a sweet scene."

Doug said most directors were happy to get one shot of the bear and then another shot of the actor responding to the bear (offscreen).

But because Bart and Freeman already knew each other, Doug protested. "I said, let's teach Bart to act that feeling of 'Do I know you?'" Doug said, demonstrating. "All those little sophisticated things have intrigue for me. Close your mouth, lower your head, peek this way. In other words, those are all little subtleties."

"They're smart," Lynne again said of the bears. "They're so intelligent."

Prior to visiting Doug and Lynne
Seus, I'd watched Bart's starring
role in Jean-Jacques Annaud's 1988
classic, *The Bear*. The film, based on
James Oliver Curwood's 1916 novel,
The Grizzly King, tells the story of
an adult bear who befriends an

Lynne Seus

orphan cub (played by a bear named Youk) as they struggle to escape
two trophy hunters. Bart's performance is a master class in acting. For
the first time—and in stark contrast to how bears were portrayed in
early American wildlife art—a grizzly was presented as a sentient being
with cognitive thought, intelligence, and complex emotions.

"Mr. Annaud has elected to see things from a mostly wordless
bear's eye view and has attempted to answer the thoughts of creatures
who, as Mr. Annaud presents it, are a lot like us," Janet Maslin wrote
in her review for the *New York Times*. "The director has, through tremendous effort, coaxed from his bear actors performances that are a
bit better than those of the film's several human players."

"Bart the grizzly, who starred in 1989's *The Bear*, deserves a best
supporting actor award for his ferocious work," added Richard Corliss
in *TIME* magazine.

"The bears are marvels," cheered *The Hollywood Reporter*, which
praised Bart's "naked honesty to its emotions."

For his role in that film, Bart earned a hefty Hollywood salary—
more than Jack Nicholson did for *Chinatown*—and, like any celebrity,
he had a backstage rider, a legally binding contract that spelled out
exactly what he needed for food and drink. During the shoot, Bart
requested six whole processed chickens and eight cans of fruit punch
each day, along with a hearty supply of fruits and vegetables.

Bart's performance was not only a masterpiece of ursine acting
but also unique in his interactions with a young cub. In the wild, adult
males will often avoid, or kill, young cubs, and two bears of such dif-
fering ages had not previously shared the screen together. However,

Bart and Youk were two consummate professionals who got along and remained true to the given circumstances of the story.

Since animal actors are considered props by the Academy of Motion Pictures Arts and Sciences, Bart wasn't eligible for an Oscar, but as a tribute to his life and career, Bart and Doug Seus were asked to join actor Mike Myers onstage at the 1998 Academy Awards to present the Oscar for Outstanding Achievement in Sound Effects Editing. Although Ernest Gold, the composer of *Exodus*, recoommended Bart for a special achievement Oscar.

"Throughout its seventy years, the Academy Awards have focused only on the achievement of people," stated host Mike Myers, decked out in a tuxedo, from the podium. "The Academy has overlooked and besmirched—and, oh yes, I said besmirched—the contributions that animals have made to motion pictures, and I'm going to do something about it right now."

The curtain to stage left of Myers opened to reveal Bart seated onstage. When Myers strolled over, Doug, also sporting a tux, handed Bart the envelope—and the bruin accepted it with a gentle bite, to rapturous applause.

Standing beside the half-ton grizzly, Myers deadpanned: "I just soiled myself."

The future looked bright for Bart, but it was later that year that the Seuses found a lump on one of his paws, and more soon appeared around his neck.

"Bart had cancer," Doug told me, his eyes misting over.

The Seuses were filled with grief, but the Alaska Department of Fish and Game called soon after. A sow had been poached, leaving two orphaned cubs. Would Doug and Lynne take them in?

"We agreed immediately," said Lynne with a smile.

Bart 2, aka Little Bart, and Honey Bump joined the Seuses' other grizzly, Tank, to carry on Bart's acting legacy and serve as ambassadors for the wildness of the planet.

The bears were immediate hits in the movie industry. Tank, a silver-tipped chocolate brown grizzly, beat out fifty other animal actors to star in *Dr. Doolittle 2*. In the film, the bruin scrubs himself in the bathtub, devours popcorn while watching TV, and toasts marshmallows with Eddie Murphy.

"Tank the Bear steals *Dr. Dolittle 2* from Eddie Murphy with clever tricks and raw animal magnetism. Tank, the eight-hundred-pound bear who plays Archie, is clearly a pro," raved *People* magazine at the time.

"There's a new Ursa Major on the Hollywood scene and his name is Tank," added Susan Wloszczyna in *USA Today*.

Steve Carr, the film's director, said of Tank, "He has actually got a lot of range. He is a bear who is committed to his art as an actor."

In addition to starring in *Dr. Doolittle 2*, Tank also sat on Jay Leno's couch during *The Tonight Show*, shared the screen with Matt Damon in *We Bought a Zoo*, and nuzzled up to Brad Pitt in a documentary about Doug and Lynne Seus, *Growing Up Grizzly*.

"Being kissed by a huge grizzly was a little unnerving," Pitt said in the documentary, "but it was an experience I'll never forget."

As Little Bart and Honey Bump grew, they joined the family business, creating a multigenerational acting dynasty that rivaled the Redgraves'. The A-list celebrities the Seuses' bears have worked with look like the cover of *Vanity Fair*'s Hollywood issue: Sean Penn, Jennifer Lopez, Burt Reynolds, Zac Efron, Sam Elliott, Steve Carell, Billy Bob Thornton, Alec Baldwin, Anthony Hopkins, Matt Damon, Emile Hirsch, and Kevin James, among many others.

"Was there one actor the bears particularly bonded with?" I asked.

"Anthony Hopkins was absolutely brilliant with Bart," said Lynne. "He acknowledged and respected him like a fellow actor. He would spend hours just looking at Bart and admiring him."

When the director of *The Edge* wanted to use a stunt double for a scene where Hopkins's character encounters Bart on a wooden bridge, the actor refused.

"Tony is the world's greatest gentleman," Lynne remarked. "He said,

'I'd really rather do it myself because the bear likes me, and I like the bear.'"

As Lynne stood to refill my coffee, a friendly guy named Smitty—the Seuses' caretaker and bear handler since 1991—poked his head in. "Honey Bump is outside if you want to see her."

Following the demonstration—during which Doug wowed me with all of Honey Bump's moves—we returned to the living room, and I accepted Lynne's offer for some elderberry pie.

I was curious if the Seuses thought there was something in particular that the bears had taught them.

"What haven't they taught us?" Doug replied quickly before growing more thoughtful. "They taught me the importance of our planet, this beautiful hanging ball. I learned that being with wild spirits increases the fever you have for the planet."

Lynne mentioned the bears taught her how to follow a dream. "To start out on our path, little by little, and to have two children at the same time, and to build it to where we can put food on the table" was a privilege, but it was even more special when they reached a point that they were able "to give back, through Vital Ground, what the grizzly bear has given to us and our family."

Vital Ground is a nonprofit land trust that Doug and Lynne founded in 1990. "First and foremost, it's an accredited land trust and our number one mission is to secure land for the grizzly bear, either holding it outright or in easement," Lynne explained. "It's not just for the bears, but they are, of course, the footprint. But also, all the plants and animals below them."

Given the large range grizzlies have, corridors are important so bears can travel to search for food and mates, increasing genetic diversity within populations. As things stand today, wild bears currently live on ecological islands, floating between private land and human development.

"We couldn't really change that Bart was born in captivity and spent his life in our human world," said Lynne. "But Vital Ground is something we could do in his name for his wild brothers."

Vital Ground first purchased 240 acres of prime bear habitat on the eastern front of the Rocky Mountains in Montana, and in the years since, the organization has conserved over one million acres of habitat.

"The permanent connection of wild strongholds is the only way to ensure the long-term recovery of grizzly bears and other native species in the Northern Rockies," said Doug.

Vital Ground either buys land outright or partners with ranchers, farmers, or private homeowners on conservation easements that voluntarily limit specific uses like subdividing to ensure that the land retains its conservation and agricultural value. The land trust also partners with local communities to reduce human-bear conflict by helping support electric fences, bear spray rentals, bear-resistant trash cans, and carcass pickup when cattle die.

Glen Willow Ranch in Choteau, Montana, on the Rocky Mountain Front is one example of Vital Ground's reach and impact. Glen Willow's owner, Mary Sexton, purchased the 630-acre farm (once owned by her grandfather) in 2014 and, five years later, partnered with Vital Ground on an agriculture land easement that ensured her property would never be commercially developed or made into dispersed residential land but would remain a wildlife sanctuary and working farm.

Every spring and fall, her property becomes a seasonal sanctuary and travel corridor as grizzly bears, deer, elk, and a whole host of other animals and birds travel the Teton River tributary from Glacier National Park and the Bob Marshall Wilderness toward Montana's central plains.

"Many feel, as I do, that we're very fortunate to live in a place where bears exist. We do have a lot of wildlife come through. Plus, it's been a productive farm ground for well over a century. I want it to remain, not only in agriculture but also as good habitat for wildlife," Sexton said at the time of the partnership.[1] "I think a conservation easement is the best protection you can give a property because in the long run we're all temporary. We won't be here. We can't determine what happens to land when we're gone, but I think Vital Ground understands [that] to have a vital grizzly bear population, we also have to have private landowners interested in maintaining the quality of their land. Working lands are good habitat."

As Doug and Lynne spoke of their conservation efforts, their cat Digger resettled on my lap, purring for more pets.

"Where the grizzly can walk, the earth is healthy and whole," Lynne declared.

By this time, the winter sun was setting behind the Wasatch Range, it was time for me to take my leave. Doug and Lynne walked me to the door, concerned that I'd get hungry on the drive back to Jackson.

"Sure we can't send you with some pie? A coffee for the road?"

"I'm fine," I replied, and thanked them for their time.

Lynne told me it was hard to sum up forty-six years of working with bears into just a few hours, so she invited me down for another visit. "And this time I promise to make you dinner."

Doug hugged me and added there was another dimension of the outdoors that you experience only in bear country. "They take the casualness out of a hike and make you aware of the entire ecosystem. Whether you love bears or just the natural world, seeing a paw print on the ground changes your day, and sometimes your life."

---

1. Mary Sexton, "Glen Willow: A Land Legacy Protected," Vital Ground Foundation, March 15, 2019.

Chapter Four

"A Most Tremendous Looking Animal"

We have passed many hours of excitement and some, perchance
of danger, in the wilder portions of our country. . . . Imagine the
startling sensations experienced on a sudden and quite unexpected
face-to-face meeting with the savage grizzly bear—the huge
shaggy monster disputing possession of the wilderness
against all comers and threatening immediate attack.

—*John James Audubon, 1848*

Peter Mangolds

The National Museum of Wildlife Art sits across from the elk refuge, two miles north of the famed antler arch in Jackson Hole's
town square. Earlier that week, I noticed an ad for an exhibit at the
National Museum of Wildlife Art in Jackson, *While They're Sleeping:
A Story of Bears*, and decided to check it out.

As I turned off Highway 191 north, I spotted a maze of animal
tracks in the snow and mule deer, grazing among the sagebrush, alongside bronze statues of bison, elk, and moose. Which animals were alive?
Which were part of the museum's famed sculpture trail? As I navigated
up the winding road, I loved that there seemed to be no distinction
between art and nature.

Founded in 1987, the National Museum of Wildlife Art occupies
a state-of-the-art, 51,000-square-foot building, inspired by the Slains
Castle in Scotland that emerges from the surrounding hillside like a
craggy rock outcropping. The museum has fourteen unique galleries
featuring over five thousand works of wildlife art from more than
five hundred artists, ranging from Native American to contemporary
masters—including Georgia O'Keeffe, Andy Warhol, Carl Rungius, and
John James Audubon.

"I'm here to see the bear exhibit," I told Sarah, the friendly clerk at
the admissions desk. After I paid my admission, Sarah handed me a
pin for my shirt collar, pointed across the room, and said, "In the King
Gallery just over there."

While They're Sleeping: A Story of Bears told a tale of bears between
1846 and 2019, with a focus on the three species of bruins in North
America—black, brown, polar—along with a special photography
tribute to local Bear 399 and her four cubs by photographer Thomas
Mangelsen.

The exhibit posed many of the same questions I'd been grappling
with: *Why do bears have such a hold on human imagination and*

fascination? How have our values and beliefs regarding bears defined and redefined them over the last 150 years? And, most importantly, *How can we coexist?*

Tammi Hanawalt, curator of art at the museum, told me that the idea for the exhibit came when one of the museum's board members attended a college reunion in Kansas and all her classmates wanted to know about 399.

"Even in Kansas, everyone was following 399's journey," Hanawalt explained. "She's really an ambassador for the species. She's symbolic of the mother bear and survival since she's raised a half dozen litters beside the road. She's so clever to have figured out that her cubs are safer and more protected from adult male bears when she keeps them closer to humans when they're young."

Because there are so many cultural stories around bears, Hanawalt said she wanted to look at how representations of them have changed over the years, along with how humans have related to them. "Since we had so many artworks that feature bears in our collection, I thought it would be a great idea to do a bear exhibit based on what we already have in the museum, to consider how bears have been depicted from the historic to the present."

When it came time to name the exhibit, which ran from October to May, a colleague of Hanawalt's remarked, "You're going to do an exhibit about bears while they're sleeping."

"I thought, *That's it! That's a great name for the exhibit,*" Hanawalt said.

As I entered the cool air, dim lights, and carpeted floor of the King Gallery, I found myself in an enchanting ursine world of elaborate paintings, stone carvings, and brass sculptures.

Peepeelee Kunilusee's *Shaman Becoming Bear* sculpture refers to the Inuit belief that animals and humans could effortlessly transform into one another. William Holbrook Beard's 1886 oil painting *So You Want to Get Married, Eh?* represents a pair of young bears who stand in front of an older-looking father-in-law bear, beseeching a nuptial blessing. Holbrook Beard often used bruins to lampoon Victorian society. In his other satirical animal paintings, Beard showed scenes of love, jealousy, and conservative Wall Street investors who created bear markets. In Robert F. Kuhn's 1987 *Clown of the Woods*, he painted a black bear reclining humorously on a rock to show the playful nature of *Ursus americanus*. And, in Robert McCauley's 2014 painting, *The Only West Left Is the One in Your Head*, he celebrated the western frontier as a vision, transformed through waves of settlement, and McCauley called bears "wayshowers." He says: "They have shown us the way through the wilderness and continue to teach us how to adapt to a changing world."

I also discovered Clifford K. Berryman's cartoon at the exhibit. *Drawing the Line in Mississippi* ran in the *Washington Post* in 1902 and led to the creation of the teddy bear.

In 1902 President Theodore Roosevelt went out to hunt black bears on a plantation with Andrew H. Longino, the governor of Mississippi. Despite being an avid hunter of buffalo, bighorn sheep, elephant, and elk, Roosevelt was unable to harvest a black bear. Frustrated, he left the tree stand and stormed off in a huff.

While he was gone, Roosevelt's hunting guide, a former slave named Holt Collier, managed to capture a young black bear and tie it to a tree for the president to shoot upon his return. But when Roosevelt saw the bear, whining and struggling to break free, he decided it would be unsportsmanlike to shoot it.

Berryman's cartoon showed the twenty-sixth president of the United States protesting with one arm stretched out straight and holding a rifle in the other as the tiny bear looked on. Later that year, Morris Michtom, a Russian-born business owner in New York City, heard the story. He created a stuffed animal he named "Teddy's Bear," and a phenomenon was born.

As I continued through the exhibit, I noticed bears seemed to embody, at once, all the powers of deities in one fascinating beast—horrifying, playful, benign, mischievous, and wrathful—and were alternatively portrayed as biology specimens, trophies, faithful moms, cultural symbols, unearthly beings, emissaries of environmental issues, and a means to express human thoughts and feelings.

The first representations of grizzly bears in North America were virtually all negative.

John Woodhouse Audubon's 1848 hand-colored lithograph, *Grizzly Bear,* which appeared in the three-volume work *The Viviparous Quadrupeds of North America*—the definitive text of nineteenth-century mammalogy—depicted one snarling bear with long claws and sharp teeth.

John Alexander Harrington Bird's early twentieth-century painting *Bison Fighting Bear* showed the two raging animals locked in mortal combat. Charles R. Knight's 1924 bronze sculpture titled *Grizzlies* depicts a raging bear, standing up and ready to attack, and Frank Tenney Johnson's ghostly painting from the same era, *Grizzly Bear,* is a haunting image of an otherworldly bruin in hushed light, certain to cause nightmares.

Before national newspapers and social media, such art was pivotal to creating public support or disdain for America's land and animals. Thomas Moran's stunning pencil and watercolor field sketches helped establish Yellowstone National Park in 1872, and Thomas Ayre's beautiful 1855 drawing *Valley of the Yosemite,* along with John Muir's majestic prose, helped create a national park in the Sierras.

When it came to grizzly bears, this early artwork fed the scary stories and superstitions surrounding brown bears. There were more comprehensive depictions of brown bears in the ensuing years—and later part of the exhibit—but as I left the museum that afternoon, I was certain that those early depictions of grizzlies had lodged in our collective unconscious and were the root of negative attitudes that still persist today.

Seeking more information, I reached out to bear expert Chris Servheen, who I'd heard speak at the Human-Bear Conflict Workshop in Lake Tahoe.

"There's an old painting called *Loops and Swift Horses Are Surer Than Lead*," Servheen began. "It's from 1916 but it typifies the attitudes about bears in general, and grizzly bears in particular, in the 1800s in the western US."

The painting is by the great cowboy artist Charlie Russell. The oil painting—which now hangs at the Amon Carter Museum of American Art in Fort Worth, Texas—depicts a team of cowboys roping a grizzly bear before dragging it to death.

Prior to the arrival of European settlers and trappers, Native Americans had coexisted alongside grizzlies for thousands of years. At the time, over a hundred thousand brown bears roamed North America (including fifty thousand in the Lower 48), from the Arctic down to Mexico, throughout the Southwest and into the Great Plains.

The first reports of brown bears by Europeans came when Sebastián Vizcaíno, a Spanish soldier, entrepreneur, and explorer, observed grizzly bears feasting on a dead whale on a beach in Monterey, California, in 1602. Henry Kelsey, an English fur trader, sailor, and the first white man to reach the Saskatchewan River in Canada from Hudson Bay, described spotting brown bears on the Canadian prairie in 1691.

The first known written record of the species was documented in 1790 by Canadian explorer and author Edward Umfreville. "Bears are three kinds: the black, the red, and the grizzle bear," he wrote, referring to the silver-gray tips on the bears' brownish fur. A few years later, Scottish explorer Sir Alexander Mackenzie wrote in his journal about an encounter with a "grisly bear" that left tracks "nine inches wide" along a stream bank.

Despite these initial accounts, most Americans learned of the grizzly bear through the accounts of Captain Meriwether Lewis and Lieutenant William Clark's expedition along the Missouri River to the Pacific Ocean.

"A most tremendous looking animal and extremely hard to kill," Lewis wrote in his journal on May 5, 1805.

At the time, bears were living along the rivers because that's where the water and the bison carcasses were. Unfortunately for them, it was also where Lewis and Clark were.

"Lewis and Clark went out of their way to shoot every bear, and in doing so, they created aggressive bears and then they described that aggression. Bears were the dominant carnivore in the landscape, and they weren't used to avoiding people, and the result was interaction," Servheen told me.

"Is that where the *horribilis* name came from?" I inquired.

"Most of it was their own making because if you shoot at a bear, and they try to attack you, he's horrible," Servheen said. "Most animals don't try to get the people that shoot at them, but grizzly bears do. They're pretty intolerant of that."

However, Servheen added, nowadays bears are pretty good at avoiding people. "That's how come they're still on the landscape."

In 1845 began the era of Manifest Destiny, a phrase coined to describe the idea that the United States was destined by God to expand its dominion and to spread democracy and capitalism across the North American continent. Nature was viewed as corrupt, and animals, especially predators, were to be conquered and controlled. As a result, bears, bison, and wolves (among countless other species) were trapped, shot, or poisoned with strychnine.

"The irony is that California thought highly enough of the grizzly bear to put one on the state flag and then, by 1924, killed every last one," Servheen said with a chuckle.

By 1930, brown bears had disappeared from Utah, Nevada, Arizona, New Mexico, and the Golden State. By the 1960s, bears, bison, and wolves were nearly extinct, and the American public rebelled.

Less than two hundred years after European settlers arrived, grizzly bears in the Lower 48 faced continuous persecution throughout most of their range, dropping from populations of fifty thousand to less than seven hundred. By the time they were added to the endangered species list in 1975, they occupied only two percent of their former range.

Things began to change in 1963. "All native animals are resources of inherent interest and value to the people of the United States. Basic governmental policy should be one of husbandry of all forms of wild-life," wrote conservationist A. Starker Leopold and a special advisory board in their influential paper officially titled "Wildlife Management in the National Parks" but colloquially known as the Leopold Report.[1]

On March 9, 1964, the US Fish and Wildlife advisory board declared at the North American Wildlife and Natural Resources Conference, "It is the unanimous opinion of the Board that (predator) control as practiced today is considerably in excess of the amount that can be justified in terms of total public interest. As a consequence, many animals which have never offended private property owners or public resources value are being killed unnecessarily."

Things began to change slowly in 1964, but bears continued to be killed right up until the 1980s, at which time human behavior began to be recognized as the root cause for most conflicts between humans and bears.

"This was a big change for animal damage control and wildlife services," Servheen said. "It switched from blaming the bears to blaming the humans. From saying, 'Bears are the problem, so we kill them,' to 'humans are the problem, so we teach them.'"

Serving as the grizzly bear recovery coordinator, Servheen spent much of the 1980s traveling all over the West and holding public meetings about conserving grizzly bears and reintroducing them in the Bitterroot ecosystem. The meetings were rowdy affairs, typically held in the evenings, after many people had downed a few drinks. People would parade around the room with "No Grizzly Bears" signs, and sometimes with a police presence.

"People's opposition was often very spirited, both to bears and to the idea of reintroducing bears. And in opposition to anybody in the audience who was pro-bears," said Servheen. "They were sometimes threatening other people in the audience who were saying it would be

---

1. A.S. Leopold et al, "Wildlife Management in National Parks," *US Fish & Wildlife Publications*, March 4, 1963.

nice to have bears on the landscape with other people in the audience standing up and yelling at them, saying, 'I'm going to kill you,' and sometimes charging individuals that were in the audience trying to testify, and once police tackling people in the aisle before they got to the stage."

A lot of the resentment was over the idea of the federal government being involved with local activities and closing the dense network of logging roads on public lands, which Servheen explained were "a huge detriment to bear survival as well as elk and other species." But the protestors feared the economic impacts of reduced timber harvests as well as the safety of their families and livestock.

There was also the difficulty of getting all the stakeholders—state, park, federal, private, and tribal—to work together across jurisdictional lines not only to address bear mortality and habitat security but also to manage trash and garbage in the frontcountry and backcountry.

Servheen quickly realized he wouldn't be successful trying to argue a position. "If people hate grizzly bears, you're not going to change their mind about that," he told me. "You're dealing with strong emotions that drive their feelings and behaviors."

The old rule about public meetings applied—no one comes because they agree with what you're proposing. So, the people at public meetings are not necessarily a representative sample of the population, yet those who spoke in opposition were usually the stories told by the newspapers, media, and reporters who attended those meetings.

"We don't want bears killing our cattle," a rancher said at the time.

"I don't want to get mauled if I go into the woods with my family to pick huckleberries," added a concerned mother in Montana.

Servheen assured them that management agencies shared the same goals. When people realized that many of their interests and concerns were shared, they started finding some common ground. In many cases, Servheen found, it was most useful to just listen and get people to talk rather than presenting to them or arguing with their positions.

"Grizzly bears are reoccupying places where they've been

exterminated for over a hundred years," he said. "Since in many of those places, people are unfamiliar with grizzly bears, their lack of experience colors their response, whereas people more familiar with bear habits know that grizzly bears aren't out to get them, and that grizzly bears don't kill all the livestock on the land."

I told Servheen about Bear 399 hanging out near my fire station, my home, and around the nearby hiking trails. "Are we having another kind of Lewis and Clark moment throughout the West because people are suddenly encountering grizzlies for the first time?" I wondered.

Servheen agreed. Thanks to dedicated work by him and many others in state and federal agencies over the years, the grizzly population has now expanded to more than two thousand bears, spread out around six key recovery zones in the contiguous United States: the Greater Yellowstone Ecosystem of northwest Wyoming, eastern Idaho, and southwest Montana; the North Continental Divide of north central Montana; the North Cascades of northwest Washington (where there are not bears anymore); the Cabinet-Yaak Ecosystem of northwest Montana and northern Idaho; the Selkirk Mountains of north Idaho and northeast Washington and the Bitterroot area of central Idaho and western Montana.

The value of grizzly bears, Servheen said, is multidimensional: "They're an iconic wilderness species. If we have wild country, then we have grizzly bears." But he pointed out that grizzlies can't easily live in a developed area with lots of people, like California, for example. "So, we need wild spaces and large amounts of wild country to have grizzly bears. And there's very little of that left in the Lower 48 states these days. So, grizzlies are an indicator species of ecosystem health and ecosystem wildness."

Brown bears also have an existential value and cultural value. Just the knowledge that bears are out in the mountains of wilderness areas like Yellowstone and Grand Teton or Glacier is important. "You may not see them, but that existence value is important," Servheen said. As is the cultural value for Native peoples, "the sense is that grizzly bears linked them to their past and to a time when they were in touch with their environment to a great degree."

At the end of our conversation, Servheen told me bears, much like great white sharks and wolves, often get a bad rap: "Most of the time, when people hear about bears, it's only because bears have done something wrong. Those are things that make the newspaper. The times bears avoided people or lived among livestock their entire lives and never killed any, none of that makes the newspaper. So, people's experience is colored by what they hear, and unfortunately, they only hear stuff that's negative most of the time."

I asked Servheen about the threats grizzlies face.

"People think they're this really massive, strong species that are able to live in difficult conditions and no matter what they'll always be there because they're so big and strong and fierce," said Servheen. "But the fact is that grizzly bears are very vulnerable to humans and human activities. So in that sense, grizzly bears are very different from what most people think of them. Instead of being an invulnerable, strong species that can live in wild country, they're a very vulnerable species whose existence is totally dependent on humans and human behavior."

I felt so thankful for my odyssey into the wild, secret life of brown bears thus far, but to answer all my questions, I knew I needed to spend more time with them in the field. I yearned to visit Brooks Falls in Katmai National Park, the site of the immensely popular Fat Bear Week competition. I dreamed of visiting Kodiak Island, home to the largest grizzly bears on the planet, and had heard rumors about a hidden, Eden-esque sanctuary where, according to legend, neither man nor bear feared the other and where it wasn't uncommon to see sixty to eighty grizzlies feasting on salmon in the river at the same time.

To journey to those places, I needed to venture to the last frontier and land of the midnight sun.

Alaska.

Chapter Five

The Island of Behemoth Bruins

Fresh bear scat on the trail, unpleasant crashing noises back in the dark of the woods and brush, reminded us that we were intruding, uninvited, into the territory of . . . GRIZ.
—*Edward Abbey*

Jennifer Smith

I followed the woman with long black hair, tattoos, an Irish accent, and a 12-gauge pump shotgun slung over her shoulder to the waiting helicopter.

Shannon Finnegan, a bear biologist on Kodiak Island, Alaska, had invited me to join her on a mission to recover GPS collars dropped by brown bears on Afognak Island in the Kodiak Archipelago, a mountainous island of mist and mossy Sitka spruce forests to the north.

"Sounds great!" I immediately answered.

Meaghan was in too.

Finnegan, twenty-nine, grew up on a small dairy farm in Sligo, on the west coast of Ireland. While other kids were watching *The Matrix* and *Twilight*, Finnegan was binge-watching all of David Attenborough's nature documentaries. After receiving a zoology degree from Liverpool John Moores University and a master's in biodiversity and conservation at Nottingham Trent University, she earned a PhD in ecology from the State University of New York–Environmental Science and Forestry (SUNY–ESF) in collaboration with the Alaska Department of Fish and Game (ADF&G) and now worked as a research biologist with Koniag, an Alaska Native corporation advocating for the Alutiiq community, the first inhabitants of Kodiak Island.

The ink on her arms was a tattooed résumé of the species with which she's worked: jaguars in Brazil, leopards in South Africa, tigers in India, and now brown bears on Alaska's Emerald Island.

"You can hop in back," Finnegan said, directing me and Meaghan to the backseat of the black Robinson R66 chopper and handing me the shotgun as she climbed in the front seat and closed the door. "You two ready?" she asked in her Sligo accent.

"Yup," I replied, half-convincingly.

When Meaghan gave me a look, I uttered the phrase all husbands say when they've dragged their wives on some mad, half-planned adventure: "Love you, honey!"

With that, our pilot—a shiny-domed, sarcastic guy named Josh—lifted the helicopter and pointed it toward Afognak, twenty-five miles to the north.

Meaghan and I had arrived a few days prior to discover everything was big on Kodiak, beginning with the island itself. At 3,595 square miles, Kodiak is the second-largest island, after the Big Island of Hawaii, in the United States. The red king crabs that frequent Kodiak's nutrient-rich, North Pacific waters can weigh up to twenty-four pounds and have a leg span of nearly five feet. Air Station Kodiak is the largest US Coast Guard base and often flies big search-and-rescue missions out to the Bering Sea. The Pacific Spaceport Complex on Kodiak's eastern coast, one of only four orbital vertical launch sites in the United States, has the largest launch azimuth range of any spaceport in the US, promising to take intrepid astronauts "from the last frontier to the final frontier."

And then there are the bears.

Standing ten feet tall on two legs and weighing up to fifteen hundred pounds, the bruins in the Kodiak Archipelago are the largest brown bears on earth. And there are a lot of them—nearly thirty-five hundred bears, one every .7 miles—which means your chances of encounter during outdoor activities are quite high.

Meaghan and I spent our first week on Kodiak bear viewing and spin fishing at a bear camp on the west side of the island. Amid a backdrop of towering peaks and fall colors, we fished for silver salmon beside grizzly prints in the gravel and bear daybeds and large belly holes dug into the sand. Whenever a bear appeared in the distance, we'd set down our fishing poles to watch them, snapping photos and observing their behavior through binoculars.

From the tidal flats, our guide, Hiram, would zoom us upriver in the flat-bottomed skiff and we'd see bears standing majestically on bluffs, digging for salmon eggs, or just cooling off in the late summer heat with a casual swim.

Despite their massive size and wide cowboy swaggers, the bears seemed to appear—and vanish—suddenly and silently. I always had the feeling of being watched, yet surprisingly I rarely felt scared. These weren't the food-stressed, hangry bears contending with millions of tourists each summer that I was used to encountering in Yellowstone or Grand Teton. The bears on Kodiak are privileged to lead a life with lots of land, few visitors, abundant vegetation, and multiple berry crops and salmon runs.

Still, when a six-hundred-pound grizzly with a patchy, tobacco-colored coat waddled toward us while we fished one day, the pucker factor was real.

Meaghan and I reeled in our lines and, as advised, huddled together in hopes of appearing bigger.

"Stay calm and stand your ground," Hiram advised, standing a few feet behind us.

"Trying," I replied.

"See his triangle-shaped ears too big for his head?" Hiram asked. "The long, lanky legs?"

"Yes."

"He's a subadult, the juvenile delinquent of the bear world. He probably got kicked out by mom this spring."

Subadult bears, typically below the age of six, are independent from their mothers but aren't sexually mature and behave differently from cubs or adults. "They're the teenagers of the bear world," Hiram explained.

Regardless of his age, the bear was five yards away from us and closing in. I reached for Meaghan's hand.

Hiram told us to observe the bear's demeanor. "He's not swaying his head back and forth, not salivating, huffing, clacking, or popping his jaws. He's not low to the ground, swatting the air, and his ears aren't back."

"So, what does that mean?" I managed.

"He's curious," Hiram said. "Subadults like to explore and test boundaries, which is why you stand your ground."

As the bear continued to approach slowly toward us, Hiram raised his voice. He had a gun and bear spray on his belt but wasn't reaching for either.

"You don't want to be here," he told the bear. "Keep moving."

The bear paused, gazing over at us, then back to Hiram. When the subadult went to take another step forward, Hiram softly clapped once, and the bear turned and scooted off like a scolded puppy.

Meaghan and I watched the bear amble away, gazing over his shoulder as he walked. He really was just like a teenager, pushing buttons and testing. When he was twenty-five yards away, he plopped down in the sand to recline against a piece of driftwood.

"That was intense!" I said, releasing Meaghan's hand.

Bears will mirror our reaction, Hiram told us. "Yelling and screaming will cause them to get anxious and stressed, and you're more likely to have an adversarial encounter." We could adjust our response, based on the bear's behavior. "If he's just curious, you don't have to start yelling and squirting bear spray all over the place like all of them floaters."

Floaters was the term Hiram gave to the weekend warriors on the river, visitors who'd get dropped off at the lake and then float down the river, camping and fishing, before getting picked up at the bay. Hiram's problem with the floaters—apart from crowding us at the fishing holes—was that most of them hadn't educated themselves about the best practices in bear country and didn't share his leave-no-trace environmental ethic. The floaters encouraged conflicts by continuing to cast when grizzlies were in the area, by displacing bears with their boats when motoring up near them, and by leaving fish entrails onshore or occasionally tossing salmon to bears as a treat. The floaters didn't have a vested interest in the landscape that a guide like Hiram did, and they didn't do their part to keep bears wild.

Along with a free helicopter tour of the archipelago, I'd initially imagined a carefree day of hiking with Finnegan. We'd get some exercise, help ADF&G get vital information, and save money by locating the lost collars, then perhaps picnic beside a majestic alpine lake.

But as Meaghan and I squeezed into the back of the helicopter that afternoon, Finnegan said we shouldn't necessarily expect the collars to have been removed. They could still be attached to the bears. "All we know is the grizzly has stopped moving," she said. "The bear could be dead, sleeping, or seated on a carcass."

We pulled on our headsets, and as we rose into the air, I told Finnegan about our various bear encounters.

"Their individualistic, charismatic behavior really fascinates me too," she replied. "No two bears are the same. They're extremely intelligent, and when you look at them, you can see they're gazing back at you and thinking the same way we are."

Despite biologists' and behaviorists' best efforts, bears continue to bust up theories and leave everyone baffled.

"One instance, we spotted two bears on Afognak, a bigger bear and a smaller one," Finnegan began. "It was June, so we thought it was a breeding pair, a large male and a smaller sow. But it turned out it was a big old boar and a male cub less than two years old hanging out together."

"I thought big males played no part in raising cubs and didn't associate with them unless they were trying to kill them?" Meaghan asked.

"Exactly!" said Finnegan. "So, things like this are always fascinating to us."

As we flew above the clear waters of Marmot Bay and approached Afognak, I spotted a pair of whales breaking the surface—their fins like the blades of two figure skaters in perfect unison—and seals clinging to tiny, rocky islands.

Finnegan told us that they'd collared 140 bears over the last four years to study seasonal movement patterns, behavior, denning and reproduction habits, and activity patterns and habitat use. Finnegan was specifically studying the bears' home range, along with dietary and energetic ecology. "I'm looking at what drives home range fluctuation," she explained. "What are the ecological aspects influencing the home ranges—when they contract, when they get bigger—plus the difference between sexes and whether they have cubs."

The collars consisted of a GPS unit, battery pack, and accelerometer, a kind of Fitbit for bears. The biggest advantage of GPS collars over the VHF collars is they can detect data remotely day or night. The GPS collars are also designed to expand as a bear puts on weight and are programmed to fall off after two years. They could also be released remotely by biologists if they determined the device was stressing the animal.

When Finnegan and the team—led by Dr. Nate Svoboda, a biologist with ADF&G I'd met at the Human-Bear Conflict Workshop—placed a collar on a tranquilized bear, they obtained hair and blood samples to assess the bear's genetics and overall health. They would also measure the bear's height and weight and extract a tooth for examination. "A premolar is very similar to a tree because it has rings, or lines of demarcation, for each year of growth," Finnegan explained.

Finnegan had observed some general trends: males had the largest ranges; their movements were highly influenced by habit diversity; and all bears had the biggest ranges during the summer months. "But sex and geographic differences exist."

There were also some movement patterns that defied all expectation or understanding, such as when a sow was observed taking her three cubs on an extended walk—and swim—across the archipelago. The four-pack not only marched across Sitkalidak Island but also swam across Ugak and Kiliuda Bays, wide bodies of water with notoriously rough seas. This wasn't a singular occurrence either. Evidently, it wasn't uncommon for fishermen in the Kodiak Archipelago to spot the grizzled head of *Ursus arctos middendorffi* bobbing in the frigid waters of the North Pacific.

Finnegan was also looking into a recent theory of trophic mismatch, which occurs when a predator fails to capitalize on a key resource pulse such as spawning salmon, causing a change in predator-prey interactions and nutrient transfer. "Kodiak Bears Found to Switch to Eating Elderberries Instead of Salmon as Climate Changes," one headline read in 2017.[1]

On Kodiak, warmer temperatures have led to elderberries ripening earlier, meaning that they now overlapped with the arrival of spawning sockeye. "This led to quite the media storm when it suggested that bears switched from salmon to elderberries when both were available to them," said Finnegan.

"Kind of like a child choosing hard candy over meat and potatoes," I suggested. "So do you believe it?"

Since the study had looked at only sockeye salmon and elderberries, even though there are many types of berries on Kodiak (elderberries, blueberries, salmonberries, and devil's club berries, among others) along with multiple types of spawning fish (coho, sockeye, and pink salmon, along with steelhead trout), Finnegan was still collecting more data before she could support or refute the claim. "By completing a stable isotope analysis, biologists are able to determine what the bear was eating, when, and the GPS collar tells them exactly where."

"How do you know when a collar's been dropped?" Meaghan asked.

Finnegan explained that collars continued to send out GPS locations once they fell off, which presents to biologists as a cluster of locations in one spot. "Again, we say it's dropped, but it could still be attached to the animal."

Finnegan suddenly stopped talking, turned, and told our pilot we were approaching the first location of a dropped collar and to start looking for a good place to land.

"Our hope is the sound of the helicopter will flush out any bears in the area," Finnegan explained, "so we don't encounter them on the ground."

The process of searching for a collar was simple on paper but more

---

1. Bob Yirka, "Kodiak Bears Found to Switch to Eating Elderberries Instead of Salmon as Climate Changes," *Phys.org*, August 22, 2017.

difficult in practice. When biologists notice a cluster of GPS locations around the same area over an extended period, they create a mean center area that they enter into a handheld GPS. During winter, they assume this cluster swarm is a bear hibernating, but if the grizzly never emerges from the den, they assume the collar has been dropped or the bear has died. Once on the ground, armed with the handheld GPS and knowledge of the mean center point, they do a general search within a hundred meters of the area.

"The GPS is not entirely accurate, so that's where it comes to trying to look at the terrain and see where the animal may have walked or if they used a tree to brush it off," Finnegan added. "If I don't see it within a few minutes, I turn on the VHF radio receiver and try to listen to find a direction of where it might be."

While some of the data could be accessed remotely, it was important to recover the collars because it allowed ADF&G to refurbish and reuse the units, saving thousands of dollars. It also gave Finnegan access to the energetic data from the accelerometer, which couldn't be retrieved remotely.

Josh spotted a small, grassy area to land and turned the helicopter. The meadow was framed by steep ridges and what appeared to be small bushes decorated in fall colors.

"Looks like great hiking down there," I exclaimed.

Finnegan and Josh laughed. "Those small bushes are head-high shrubs that are virtually impossible to hike through."

Josh landed the helicopter and cut the motor. When the blades stopped, we hopped out into a marshy bog, and water immediately filled my boots. The surrounding forest was primeval, silent, and had a *Jurassic Park* feel.

I handed Finnegan the shotgun as Josh slid a Glock 10mm handgun into his chest holster.

"Kodiak is big boy country," he said with a smirk.

With neither a gun nor bear spray, I felt naked and vowed that Meaghan and I should hang close to Finnegan or Josh.

"We'll start our search a hundred yards south," said Finnegan, pointing toward a steep hillside.

After ten minutes of slogging, we started up a steep ridge. Finnegan led the way, followed by Josh, Meaghan, and me. After only a few steps, it was evident this would be no easy saunter through the woods. Scaling the muddy, steep hillside through a tangled, nearly impenetrable mass of shrubs and fallen timber required an army crawl on all fours. Sweat soaked my shirt, and as my foot slipped on a wet branch, I started to slide down the muddy hill. Instinctively, I reached out for a woody stem to stop my fall. Pain shot through both hands as prickly spines pierced my palms.

Devil's club (also called devil's walking stick) is a shrub with big leaves, a noxious pyramid of red berries at the center, and prickly spines covering every part of the plant, except the berries and roots. The spines are so sharp that Alaskan Natives used to employ them as fish hooks and lures.

Along with devil's club, we also had to contend with thick stands of alder, a shrub that can reach heights of twenty feet.

"Bear heaven. Alder hell," Josh deadpanned.

Finnegan handed me her shotgun so she could focus on her handheld GPS, but now I was struggling to hike with only three extremities. After forty-five minutes of crawling on our hands and knees, sweating and swearing, bloody scratches covered all our arms and legs.

Getting through Afognak was the toughest hike I'd done in my life, including multiple twenty-four-day treks in the Himalayas above seventeen thousand, five hundred feet.

"This is crazy," said Meaghan. "I'm exhausted."

"Don't think we're going to find this one," Finnegan said. "Let's go find another."

"Thank God," I replied.

"Get to the choppa!" said Josh, mimicking Arnold Schwarzenegger in the movie *Predator.*

Once back in the air, we flew to three more collar search areas—all located on heavily vegetated cliffs.

"I can't believe bears can access these areas," I exclaimed.

"No way we're getting these collars unless we rappel in," added Finnegan glumly.

By then, we'd been out searching for nearly three hours, and bad weather was moving in.

"We'll probably have to come back in the spring when the vegetation's not so thick," said Finnegan.

"We should keep an eye on this fog," said Josh. "Don't want to get stuck out here."

After another search on a forested ridge proved unsuccessful, we landed for a third and final time. By then our arms and legs were covered with a kind of red graffiti, and the devil's club was having a systemic effect. I felt nauseated and feared I was about to get sick.

"Thirty minutes on the ground, and then we have to get back," said Josh as we hopped out of the helicopter for the third time.

"We got this!" said Meaghan with a renewed sense of purpose.

Once again, we plunged into the devil's club and alder brambles. There was such a tangle of wet branches underfoot that my hiking boots sometimes didn't even touch the earth.

Suddenly Finnegan let out an excited yelp. "Found it!" she said, holding up a worn GPS collar.

"Success!" I exclaimed.

"We can go home now," added Josh.

I was equally as excited about getting out of this jungle hell as I was about finding the collar.

As we flew home in the blowing fog, I decided that the day's adventure would be classified as type II on the fun scale: difficult and painful at the time but rewarding and fun in hindsight.

Meaghan and I were thrilled to have played a small part in assisting with finding the collar, helping Finnegan to recover her data, and saving ADF&G some money.

Kevin Grange

The day ended with all of us going to the Chart Room, the fancy restaurant at Kodiak's Best Western, serving Alaska favorites like grilled halibut and king crab. As I ate, Finnegan told us about an initiative by ADF&G and the Koniag Corporation to bring bear education to rural communities. There were six Native villages spread out across the archipelago, accessible only by plane or boat.

"We'll be flying to Akhiok, the most remote village on Kodiak in May. You should join," said Finnegan.

At the time, none of us knew our trip would coincide with the arrival of a massive bear in the village, wreaking havoc and killing pets. It would be either the worst time for us to teach a class on bears or the best.

Chapter Six

A Tale of Two Bear Parks

There are those who can live without wild things, and those who cannot.

—*Aldo Leopold*

Randy Gravatt

71

As I boarded the white safari overland truck with a green sun canopy, Nadine, a teenager wearing blue jeans and a red Yellowstone Bear World T-shirt, handed me a gray plastic tray filled with sliced bread. "You can stand in back," she said with a friendly smile, "and we'll get going soon."

The truck quickly filled with families, tourists who, like me, had signed up for the "wildlife excursion" that promised to get us up close and personal with Rocky Mountain elk, American bison, black bears, and grizzly bears.

When we were all standing shoulder to shoulder—each with our tray of treats—the tour began, and Nadine updated us on the rules.

"Absolutely no throwing food while the truck is moving. If you do, I will yell at you, and you don't want that."

Nadine also advised us to hold on to our personal items that were near and dear to our hearts: "Hats, sunglasses, cell phones, spouses, children."

If any of these items went over the side of the truck, we wouldn't be able to hop off the truck to retrieve them. "We can get them later, but that might be after a curious bear has played with them," Nadine warned.

Yellowstone Bear World is a 125-acre drive-through wildlife park near Rexburg, Idaho, which also has a mini-Ferris wheel and log roller coaster, a petting zoo, and a gift shop that serves up a colorful assortment of taffy.

I'd arrived on a sunny morning in May, signing up for a "once in a lifetime" experience to bottle-feed black bear cubs. The cubs of the year (also called *coys*) had been born in late January. Since emerging from the den, they'd been paraded around to local news stations, outdoor stores, and hunting expos. They'd also been trained to sit on their assigned wood pallets, to be fed by Bear World staff and tourists.

"Meet Emmie," an employee named Alicia said, beckoning me over and handing me a bottle filled with goat milk. "She needs lots of love because she's a princess and always gets what she wants."

Emmie was a four-month-old cub with wide eyes, short and rounded ears, and a grayish snout. Alicia instructed me to hold the bottle in my left hand and to pet Emmie's head with my right. "Just long, smooth strokes, please," she advised. "No scratching like a dog, because that initiates play, and she is here to eat."

I reached across the fence separating me from the tiny bruin. Her fur felt equal parts beagle and Brillo pad. As I placed the bottle's nipple in her mouth, she immediately began slurping up the tasty milk.

As I fed Emmie, another teenage employee, a boy with a constellation of acne across his face, held up a large camera and told me to look over at him and say, "cheese."

"Cheese!" I said as the camera flashed.

After feeding Emmie, I also fed other tiny bears with names like Ernie, Wade, and Rocky. The cubs were cute, the staff was nice, and the grounds clean and immaculately manicured. Even so, I couldn't pinpoint exactly what it was that felt off to me about Yellowstone Bear World until my friend Kendall, a former intern at Bear World, later joined me on the wildlife excursion with her two-year-old daughter.

"Bear World represents itself as a wildlife sanctuary, but none of the animals are rescues," whispered Kendall, as we tossed pellets to deer and elk that seemed only mildly interested. "It's more like a puppy mill for bears."

Kendall had interned at Bear World over a decade ago. "I was trying to get my foot in the door as an animal keeper," she explained as we approached the bear exhibit. "They sent a rep to the job fair at Colorado State University and advertised it like it was a legit thing."

As an intern, Kendall had shared a six-bedroom, three-bath home with eight other girls. She'd worked twelve to fourteen hours a day, five and a half days a week, and was paid $500 a month—$200 of which was deducted from her paycheck for housing, netting her only $300 a month for working around two hundred and sixty hours. "You do the math," she said. "My dad was really angry when he found out I took the job."

Despite its pristine appearance, Kendall said that in her experience, Yellowstone Bear World had been quite chaotic behind the scenes. There was a wolf that escaped; a grizzly bear that fled the park, only to be led back with a trail of marshmallows; black bears that climbed atop cars; and an elk that had shish-kabobed a Prius. Employees were asked to feed a large grizzly bear named Corky by hand and were put in the black bear exhibit with over sixty bruins with only PVC pipe and bear spray for safety.

According to Kendall, the breeding program was also less than ideal. "There was no DNA testing, routine veterinary checks, or ear tags to differentiate who was who. Inbreeding was rampant," she said. "We just put them all together and let them have at it. One grizzly mated with his mother but thankfully didn't produce cubs."

Along with being prematurely weaned from their mothers, the cubs were also kept on the bottle much longer than necessary in order to maximize the bottle-feeding profits from the public. There was also the matter of diet. "I'm talking about a box truck filled with Ho-Hos and moldy Wonder Bread," she explained. "It was a lot of simple carbs and sugars." As for the grizzlies, they received ground sausage and hamburger. "But they should've been getting the whole animal. They weren't receiving any organ meat and the kind of balanced diet you'd get from a whole carcass."

Kendall's experiences were recently confirmed by an undercover investigation conducted by an animal rights group and by both state and federal citations that Yellowstone Bear World has received.[1] According to the *Post Register,* Yellowstone Bear World was fined $8,928 by OSHA (Occupational Safety and Health Administration) in 2022 for putting its employees at risk of potential bear attacks and

---

1. Johnathan Hogan, "Animal activists put spotlight on Yellowstone Bear World," *Post Register,* October 10, 2021.

exposing them to a hazardous material (i.e., bear spray) without proper training. They later settled for $6,250.[2]

In 2022, the facility also received notices of violations from Idaho Fish and Game (IDF&G), stating the park wasn't providing annual veterinarian checks or documentation about animals being sold or animal deaths. IDF&G also accused the park of breaking state code by allowing the public to feed captive wildlife with its black bear cub bottle-feeding experience.

I wanted to believe Yellowstone Bear World would improve its animal care after these violations, but in 2023—as reported by the *Idaho Statesman*—a lobbyist for Bear World wrote a bill that would lessen oversight of the park and exclude them and other USDA-licensed animal exhibitors from Idaho Fish and Game oversight regarding captive wild animals.[3] Yellowstone Bear World operates under a USDA class C license, and under Idaho Senate Bill 1084—which Governor Brad Little signed into law in April 2023, IDF&G was removed from oversight on class C exhibitors. Regulatory duties now fall entirely on the USDA.[4]

Under the new law, Yellowstone Bear World doesn't have to provide IDF&G with birth, death, veterinary, or transfer records, nor do they need an insurance bond to cover the costs associated with any animals that escape, spread disease, or experience any other critical event.[5]

"This bill will allow Yellowstone Bear World to run its business without constantly looking over its shoulder," a lobbyist from the wildlife park told the Idaho Senate Resources Committee.

The class C license has lower standards than a zoo and allows the facility to not only exhibit animals but to buy and sell them. What bothered Kendall most was that Bear World keeps only six cubs per

---

2. Johnathan Hogan, "OSHA opens inquiry into Bear World after undercover investigation by PETA," *Post Register*, July 27, 2022.

3. Johnathan Hogan, "Idaho Senate passes bill to reduce regulations on Bear World," *Post Register*, March 13, 2023.

4. Nicole Blanchard, "Yellowstone Bear World, fined by OSHA lobbies for bill to nix oversight of wildlife parks," *Idaho Statesman*, March 13, 2023.

5. Bryan Clark, "Bear World should not be exempt from Idaho regulations," *Idaho Statesman*, March 15, 2023.

year yet produced more than that and downplayed it to the staff. "When you don't have individual records," she said, "there's no proof the bears disappeared."

"The cubs here were born in, and will remain in, our park for the duration of their lives," Yellowstone Bear World posted on its Facebook page, but the *Post Register* discovered that transfer records reveal the facility has sold ninety-six bears to similar roadside zoos since 2009.[6]

For instance, records showed that Yellowstone Bear World sold black bear cubs to Joseph Maldonado-Passage—of the *Tiger King* Netflix series—whose Greater Wynnewood Exotic Animal Park in Oklahoma was closed in 2018 after seventeen accounts of animal abuse. Maldonado-Passage is currently serving twenty-two years in prison on animal abuse charges—and two counts of attempted murder-for-hire.[7]

The *Post Register*'s research revealed more than eighty-four black bears were also found to have been sold to Gregg Woody, who ran a traveling exotic and farm animal exhibit known as Woody's Menagerie in Mulberry Grove, Illinois. In 2016, Woody's Menagerie was fined by the USDA for poor veterinary care, inadequate space for bears and big cats, and not providing adequate shelter in winter.

Another thirty-plus bruins were sold to Bear Country USA, a family business near Rapid City, South Dakota, where two members of the family, Kevin and Brendon Casey, pleaded guilty in 2006 to illegally purchasing and transporting two grizzlies from Minnesota, as well as slaughtering black bears and selling their body parts on the black market. The Caseys had earned over $26,000 by selling forty-four bear paws, twelve pounds of bear meat, and sixty-five frozen gallbladders from bears.

"Why gallbladders?" I asked.

6. Johnathan Hogan, "Animal activists put spotlight on Yellowstone Bear World," *Post Register*, October 10, 2021.

7. Johnathan Hogan, "Animal activists put spotlight on Yellowstone Bear World," *Post Register*, October 10, 2021.

"Bear bile," Kendall explained. "It's used in traditional Asian medicine and is more valuable per ounce than gold or cocaine."

Bears are the only mammals that produce the highly coveted ursodeoxycholic acid in large quantities.[8] Bear bile has been used for thousands of years to treat everything from gallstones to liver disease, heart problems, sore throats, epilepsy, headaches, and hemorrhoids. To obtain the bile, bears in places like China, Vietnam, Laos, Burma, and Myanmar are either killed in the wild or kept by the thousands in small crush cages where they can't sit, stand, or turn around and are often left dehydrated and underfed. The process of harvesting the bile in these bear farms is often cruel and painful, requiring either an inserted catheter that is highly prone to infection or creating a permanent fistula, or opening, and allowing the bile to drain.

"Mainly Asiatic black bears," Kendall explained, "but occasionally the sun bear or grizzly."

As the safari truck cleared the double gates, leading to the bear area, we passed three grizzlies lounging in the shade around a pond.

"I'll give Bear World some credit. The grizzlies do have a decent habitat with trees, water, and room to roam," Kendall said. "They probably have more space here than many zoos."

"But from the looks of it, they don't get much enrichment," I replied, noting the animals appeared listless and lazy.

As the safari truck proceeded into the black bear exhibit, a swarm of *Ursus americanus* immediately gathered around us. Nadine advised us to break the bread up into pieces no smaller than an Oreo cookie. "Remember, we're feeding bears, not tiny birds."

I reached into my tray and tossed pieces of Wonder Bread into the pack of bruins. They panhandled according to their various personalities: some stood on their hind legs, begging with their mouths open; some sat on their haunches and politely gazed up with pleading eyes; and others reached out to intercept my errant throws.

---

8. Rachel Fobar, "Bear bile facts and information," *National Geographic*, February 25, 2019.

I had to admit, it was fun watching the begging bruins—but I sensed it was sending the wrong message. Although Yellowstone Bear World is specifically a drive-through wildlife park, I couldn't help but feel concerned that feeding bears like this to get a good picture might encourage some of the tourists to act the same with wild bears in places like Yellowstone National Park, located just ninety minutes up the road.

Kendall agreed. "The message is, if you want a photo, approach and feed the bears."

Despite the sunshine and warm temperature as I drove through Island Park, Idaho, later that afternoon, I was feeling pretty depressed about the plight of bears in the world. Island Park—a collection of restaurants and gas stations, surrounded by the Targhee National Forest—was known for fly-fishing in the summer, snowmobiling in the winter, and the illegal killing of grizzlies in the spring and fall.

Most poaching incidents in this area weren't related to bear bile or parts—they were vandal killings, and sadly, Island Park was a hotspot.

"Vandal killings are when people shoot animals like bears because they don't like them," brown bear expert Dr. Chris Servheen told me.

"You mean people who believe 'the only good bear is a dead bear' and that the best way to deal with grizzlies is to 'shoot, shovel, and shut up'?" I asked, quoting a long-heard rural attitude about dealing with unwanted animals. The idea is you put a bullet in a bear, bury it, and close your mouth.

"Exactly," Servheen replied. "An unfortunate number of these incidents occur because it's easy to kill an animal in the backcountry and just let it lay there and not speak about it."

Four brown bears had recently been found poached in Island Park, only a short distance from Yellowstone National Park and with a population of less than two hundred people.

The first incident had occurred in October 2020, when a twenty-year-old male grizzly was found dead, shot broadside with a high-powered rifle in Coyote Meadows. Another dead grizzly was discovered one month later and hit broadside near Cold Springs Road. The location of the bullets was telling: a charging bear will often be hit frontside. A bear hit broadside meant more likely it was ambling away or foraging on plants or a carcass.

The third vandal killing in the span of just six months occurred in March 2021.

"Who's killing the grizzlies of Fremont County?" investigative reporter Natalie Schachar questioned in a bombshell article for the *Washington Post.*

"Is there a serial grizzly bear poacher on the loose in Idaho?" asked *USA Today.*

In March, Idaho wildlife officials had been conducting a routine flyover above Island Park when they detected a mortality signal on the collar they'd placed on a female bear. On March 14, a trail camera had captured the sow leaving her den with her cub, but there had been no sign of return. When biologists trudged through the snow to investigate, they found the sow's body half-submerged in a creek near Pole Bridge Campground, her majestic hide riddled with bullets. Her cub was found in the den, deceased, a victim of hunger and hypothermia.

"I don't think anyone is going to claim that this is hunting," Kit Fischer, the director of wildlife programs at the National Wildlife Federation, told *East Idaho News.*[9] "That's the point to make clear, this is not hunting, this is poaching. I think any sportsman will see it as that violates all the tenets of wildlife management."

"The loss of a reproductive female grizzly is a real tragedy," added US Fish and Wildlife Conservation Officer Doug Peterson.[10] "Someone

---

9. Kit Fischer, as quoted in Jeannette Boner, "$40,000 reward offered for info about illegal shooting of momma grizzly bear," *East Idaho News,* April 28, 2021.
10. Doug Peterson, as quoted in Idaho Fish and Game Press Release, "F&G seeks information regarding the killing of a female grizzly in Fremont County," April 26, 2021.

out there knows what happened to this bear and we are asking them to come forward and share that information with us."

The unlawful killing, possessing, or wasting of wildlife is a federal offense, especially in the case of a grizzly bear, an endangered species. A $40,000 reward for information was offered.

IDF&G couldn't catch the first two poachers but suspected those responsible for the death of the sow and cub were hunters with a late-season elk tag. Their investigation that March was hampered by cold, rainy weather, poor access to the site, and the fact that the carcass had already been fed on by scavengers. But a break came a few weeks later in May 2021, when officials discovered a key piece of evidence. The poachers had left behind not only a dead bear but a trail of cell phone pings that pinpointed their location.

IDF&G sent a warrant to Google to learn which devices had been in the Little Warm River area around March 14 and were led to Jerad Baum and his father, Rex Baum, who lived in nearby Ashton, Idaho. While the Baums had no connection to the two bears killed the previous fall, they confessed to having shot the sow over ten times in March, leaving her in the river, and tossing their Ruger 57 in the river to hide the evidence.

Rex Baum was ordered to spend three days in jail and was banned from hunting for ten years. After pleading guilty, Jerad Baum was sentenced to thirty days in jail, required to pay over $12,000 in fines, and was forbidden from hunting for life.

While justice was served to the Baums, many poachers in Montana, Idaho, and Wyoming claim self-defense or mistaken identity and aren't prosecuted for violating the Endangered Species Act (ESA).[11] Between 2015 and 2022, twelve hunters confessed to covering up their kills, dismembering grizzlies for body parts, and stalking the bruins, despite pleading self-defense as a reason for shooting the bears. Sadly, only one

---

11. Ryan Devereaux et al., "The Grizzly Files: Grizzly Bear Poachers Flout the Endangered Species Act and Get Away With it," *High Country News*, December 20, 2023.

of these cases resulted in prison time, and it's doubtless there were more grizzlies killed illegally that were never reported or found. Additionally, during the same period, federal officials neglected to prosecute over eighteen cases of hunters who claimed they mistook a grizzly for a black bear when they fired.

If grizzlies are delisted and without threat of prosecution under the ESA, Servheen and others fear such vandal killings would increase.

Randy Gravatt

I pulled up to the Grizzly and Wolf Discovery Center in West Yellowstone that afternoon in a foul mood. The plight of bears in general, and grizzlies in particular, seemed like such an uphill battle.

But I was eager to experience a different type of bear park. I was here at the Grizzly and Wolf Discovery Center (GWDC) to observe a live bear product test, pitting a grizzly bear against a commercial cooler. Would the cooler, and the delectable treats inside, withstand biting and thrashing by the bruin and achieve the coveted "bear resistant" stamp of approval from the Interagency Grizzly Bear Committee (IGBC), a team of biologists and scientists that handles the long-term monitoring and research of grizzlies in Wyoming, Montana, Idaho, Alaska, and Canada? Or would the product fail and the bear access the food inside? I parked and hopped out of my truck, keen to find out.

The GWDC is a nonprofit wildlife park for bears that have been removed from the wild, rescuing them from being euthanized. Founded in 1993 by a local developer, Lewis S. Robinson, to increase the understanding and appreciation of brown bears, the center has since added wolves, birds of prey, and an underwater otter exhibit. For the past twenty-one years, the GWDC has been accredited by the Association of Zoos & Aquariums (AZA), an independent organization that assesses animal welfare, veterinary care, keeper training, safety, conservation efforts, and educational programs. Maintaining an AZA accreditation puts the GWDC in the company of world-renowned facilities like the San Diego Zoo, Monterey Bay Aquarium, and Woodland Park Zoo (in Seattle).

I followed a line of tourists toward the GWDC building, passing a bronze statue of a brown bear leaping for a salmon, and gave my name at the front.

As I waited for Randy Gravatt, who runs product testing at the GWDC, I spotted numerous plaques lining the walls. The first award was from the Interagency Grizzly Bear Committee (IGBC), made up of wildlife agencies in Montana, Idaho, Wyoming, Washington, and British Columbia, plus the US Forest Service, National Park Service, Bureau of Land Management, and US Geological Survey: *From your director to your four-legged testing crew members, your professionalism and integrity have been instrumental in advancing our mutual efforts to provide and promote the use of effective bear-resistant containers that help keep both bears and people safe. Thank you all!*

The GWDC's product testing program had also won an AZA North American Conservation Award.

"It's a huge deal in the Zoo world," GWDC director John Heine said at the time.[12] "It's a very coveted award. It shows we're trying to make a difference with the wild counterparts. You have to do something tangible to give back to be recognized by AZA for that conservation work. Bears can get into trouble. By providing this service, we're able to help bears in the wild."

Gravatt, a sixty-year-old man in a ball cap and GWDC T-shirt with a walkie-talkie affixed to the back pocket of his gray Carhartt pants, approached me and extended a hand.

"Nice to meet you," he said. "Let's show you Easy Street, and then we'll start testing."

Easy Street was a Disney-like mockup of a small neighborhood, where everything was set up in a manner to entice bears to visit and steal food. Bird feeders hung low from the roof of a wood cabin, bags of dog food and pet bowls decorated the porch, and a heaping trash can with a cheap lid barely attached was certain to be a target for any wandering bruin. Lush gardens, an aviary, and a chicken coop—none with any fencing—abutted the sides of the house.

"After visiting Easy Street," a sign announced, "you realize that humans are a pretty good source of food."

"This is a reminder for us all to do our part and don't let our neighborhoods become Easy Street," said Gravatt as he led me to a graveyard for products that had failed the bear test over the years. Coolers, trash cans, dumpsters, and bear canisters in various designs and colors were all busted, bent, bit, broken, and dotted with ursine punctures.

"That dumpster weighs seven hundred and fifty pounds," said Gravatt, pointing, "and our bear, Sam, rolled it over five times."

The live product testing versus captive grizzly bear program was created out of a need to help keep bears wild and far away from human food and garbage. According to the center's website, "the protocol for testing was developed through an interagency collaboration effort

---

12. John Heine, as quoted in Mike Moore, "GWDC recognized with national award," *Bozeman Daily Chronicle*, October 2, 2015.

between the Montana Department of Fish, Wildlife & Parks, the GWDC, and IGBC."

Between April and September, the bears at the GWDC tested around sixty products.

"The manufacturer sends us a product and then we bait it," Gravatt told me as he used a spatula to apply peanut butter in long, painter-like brushstrokes to the inner compartment of a cooler.

"How long do the bears have to try to get in?" I asked.

"Sixty minutes, and the test is also videotaped," Gravatt replied, tossing in some big dog biscuits and large chunks of half-frozen rainbow trout before securing the cooler with two supposedly bear-resistant locks. "The bear has to be actively trying to access the interior, so we stop the timer if he loses interest, starts playing with the product, or drops it into the ponds."

Before-and-after photos of the product are taken, and the tests are filmed by volunteers so the companies that submit can see where, and how, their products held up or failed.

"Products usually fail around the seams and latches," Gravatt explained. "But around 50 percent pass the test."

The rules are strict: a cooler fails if a bear manages to put a quarter-inch hole—about the size of a canine tooth—or larger into the interior compartment or if the grizzly makes a one-inch opening on a trash can and the latch still has to work properly after the 60 minutes of active contact time.

"If the bear is unable to access the inner compartment after sixty minutes, the product has passed and the company can use the IGBC bear resistant logo," Gravatt added.

No products are labeled "bear proof." Gravatt told me that, given enough time, a bear can get into anything.

I'd once seen a video of a bear trying to bite an electrified deer carcass. "After getting shocked, he was smart enough to unplug the cord and disconnect the animal from the electricity," I said.

"They're very smart and resourceful," agreed Gravatt. "And they have the ability to use tools."

The yard, or habitat, where the grizzlies roamed looked like a landscape from Yellowstone or Katmai National Park: an alpine meadow with rocks, trees, and deadfall, along with a pond and stream where the bruins fished for live trout to the delight of tourists.

Gravatt placed the baited cooler onto the landscape, and I took my spot on an elevated viewing area next to the two volunteers—a middle-aged woman and her mom—recording the test.

It was a classic David versus Goliath matchup. In one corner sat a twenty-quart, hard-sided green cooler, weighing in at nineteen pounds and, according to its manufacturer, capable of twenty-four cans of beer or sixty-five pounds of ice—perfect for work, hunting, fishing, and days at the beach.

On the other side of the ring stood Seeley, a six-hundred-pound dark cinnamon brown bear with a reputation for being able to reverse-Houdini her way into any backpacking canister, small storage box, pannier, cooler, dry box, food storage locker, residential garbage cart, or dumpster.

Seeley and her sister, Condi, had both been born in the Seeley-Swan Valley of western Montana to a mother who'd been euthanized due to destructive behavior, which she was also teaching to her offspring. Once orphaned, Seeley and Condi were moved to the Grizzly and Wolf Discovery Center.

As Seeley was released into the yard, she immediately sought out the enrichment treats (i.e., meat and fish) she knew would be hidden under rocks and logs in the landscape. These snacks offered tasty calories with minimal effort, so it was only natural she prioritized them. Once finished, she locked in on the cooler, doing everything in her ursid power to access the inner compartment. She bit, chewed, clawed, rocked, and scraped at the cooler. My favorite move was when Seeley stood on all fours, then used her front paws and bodyweight to press down on the cooler rhythmically, as if delivering high-quality chest compressions.

"That's the CPR technique," said Gravatt with a chuckle.

Like firefighters struggling to extract a patient from a crashed vehicle, Seely seemed to intuitively know to work the seams, search for pinch points, and use her claws and "jaws of life" to bend the cooler and create gaps. "We do the same thing on the scene of a car accident," I remarked.

As Seeley broke through the cooler's outer layer, orange foam insulation poured out on the grass.

"We call that a spill on aisle four," Gravatt explained. "But she still hasn't accessed the inner compartment, so the product hasn't failed yet."

While Seeley continued to attack the cooler, I wandered over to speak with Tut Fuentevilla, a naturalist and curator for the Education Department at the GWDC. His job was to interact with the public, manage the naturalists, and develop creative messaging for the facility.

"The bear-resistant product test is entertaining, and it also teaches a valuable lesson," Fuentevilla told me. "Keeping bears out of human food and garbage is the best way to ensure their survival in the wild."

I couldn't help but note how the emphasis at GWDC was, as opposed to Yellowstone Bear World, entirely on not feeding bears.

Fuentevilla, a smart, athletic thirty-six-year-old who enjoyed trail running (with bear spray) and cross-country skiing on his days off, had been born in Bethesda, Maryland, and studied science and ecology at the University of Maryland before attending graduate school at Portland State University in Oregon.

Along with brown bear conservation and product testing, Fuentevilla said the GWDC aims to emphasize the importance apex predators like bears and wolves have on large-scale landscapes.

"Removing predators from a landscape is like taking a cog out of a system," he explained. "Everything starts to fail and go sideways."

Bears and wolves regulate lower species in the food chain by affecting the distribution, abundance, and diversity of their prey, an ecological effect known as trophic cascade.

For instance, although many people believe the best way to improve hunting species like elk and caribou is to remove apex predators, the opposite is true. Fuentevilla related the reintroduction of wolves to Yellowstone in 1995. "Prior to that, the elk herd would have spectacular growth, followed by huge population crashes, and no amount of hunting ever prevented that," he said. "Humans don't replace natural predators very well, and we're not able to do it in the same way as a wild predator like a bear or wolf would do it."

Since the reintroduction of wolves to Yellowstone for the past thirty years, the elk population has been very stable compared to historic population trends. "Certain animals like elk are adapted to breed in the face of predatory pressure," Fuentevilla explained. "Bears and wolves want to make it easier on themselves while hunting, so they'll target weak or sick animals and by doing so help remove illness and disease from the herd."

Furthermore, my own research showed how predators also helped build plant, bird, and mammal communities. Normally, ecosystems are looked at from the bottom-up, but as reported in *Nature Ecology & Evolution*, scientists taking a top-down approach now appreciated that predators were as influential on a landscape as photosynthesis. At Yellowstone, predators like wolves and bears keep elk populations in check and on the move, allowing more grasses and shrubs to grow beside the park's rivers and streams, resulting in cooler water, which in turn supports aquatic life such as trout. Since wolves also keep the coyote population balanced, it permits animals like pronghorn and smaller mammals to thrive, while the carrion that predators leave behind feeds scavengers like eagles, falcons, and osprey.

Fuentevilla reflected on his favorite part of the job. "Getting a window into their lives and thoughts is tremendously rewarding and surprising, and it broadens my own experience and makes the natural world richer and more impressive to observe. It's a glimpse into the life of these big, charismatic predators that would be impossible in other contexts, because they're not under threat of failing to get food and survive here," he said.

Visitors to the GWDC, Fuentevilla said, often leave with a different impression of bears than what they expected when they arrived. While many people arrive thinking of bears as only big predators and wanting to see that fierceness, after spending time at the center watching the grizzlies, they leave understanding that bears are not mindless killing machines but very complex animals. For instance, Fuentevilla said that, despite bears being very capable hunters, their diet is mainly plants, berries, and salmon because those are the foods that are usually going to be abundant in a predictable way.

Fuentevilla's main concern was educating people to change their own behaviors when it came to bear safety—for the benefit of both species. "The leading cause of death for wild bears outside of national parks is interactions with people rather than natural causes," he pointed out. "We want to educate the public about the difference between a grizzly moving through your backyard and a bear staying there because of garbage or an unsecured food source and becoming a problem."

I thanked Fuentevilla for his time and hurried back to catch the end of Seeley's live bear test against the green cooler.

"How's it going?" I asked Gravatt.

"Take a look," he replied.

Seeley had clearly dominated her competition: orange chunks of insulation foam were scattered everywhere, and the bruin had accessed the inner compartment for a food reward.

"We're at sixty minutes so we'll end the test now," Gravatt said as the volunteers stopped filming.

In the yard, Seeley stood triumphantly atop the remnants of the cooler.

"The cooler failed because the bear penetrated the interior compartment," Gravatt said later as I helped him to prep the next cooler, this one a forty-five-quart, hard-sided white cooler that held forty-five beers or sixty-five pounds of ice and promised to keep its contents ice-cold in sweltering conditions. "But companies can always redesign and retest."

As we worked, a pickup pulled up carrying two pallets of expired strawberries from Costco in Bozeman. The animals at the GWDC were fed lots of fruit and vegetables that are donated from local supermarkets, Gravatt told me. "Along with donated meat and carcasses from game processing facilities in Bozeman. And a fish hatchery in Ennis, Montana, donates over three thousand fish at the end of their breeding cycles each year."

Some of the bears at GWDC had developed testing fatigue due to the avalanche of products that arrived each spring, so the facility and IGBC had partnered with the Washington State University Bear Center to hold the same testing program with a different set of bears.

Gravatt encouraged me to visit the Bear Center. "They're doing great things there."

I had no idea a major university had grizzlies on campus, and I vowed to visit.

Chapter Seven

Tracks of the Craighead Brothers

Alive, the grizzly is a symbol of freedom and understanding—a sign that man can learn to conserve what is left of the earth. Extinct, it will be another fading testimony of things man should have learned about but was too preoccupied with himself to notice.

—*Frank Craighead*

Peter Mangolds

Seated on a steep ridge in a subalpine forest of lodgepole pine, Engelmann spruce, fir, and whitebark pine, Kerry Gunther—bear management biologist at Yellowstone—lowered his binoculars and grew reflective.

"With a lot of species like elk and bison, they all seem to do the same thing, whereas with bears, they're all individuals. They each have their own personality. And then hibernation—five months without eating or drinking. That alone is an amazing way to deal with a food shortage by sleeping."

"So they don't hibernate because of the cold?" I asked.

Gunther shook his head no and proceeded to tell me about the thick fur of bears and Bergmann's rule regarding size and shape. In 1847 Carl Bergmann, a German biologist, discovered that colder climates spawned bigger and rounder bodies, helping animals (like grizzlies) retain heat. Bergmann's rule is part of the reason coastal brown bears are so much bigger than grizzlies found in interior North America and why a polar bear is over twice as tall at the shoulder as a sun bear found near the equator.

Turned out the Arctic ground squirrel's (true hibernators) body temperature drops from 99°F to as cold as -27°F during hibernation, whereas the body temperature of a grizzly dips only 8 to 12 degrees— from 100°F to 88°F. "You could pull out a squirrel while its hibernating and use it as a football, and it's going to take a while to wake up," Gunther said. "But a bear can be up and about in just a few minutes."

"Good to know," I said.

Gunther is a bright, energetic, and immensely friendly man in his early sixties, dressed in the iconic green and gray uniform of the National Park Service. He earned a master's of science in fish and wildlife management from Montana State University and, for the last forty-one years, has worked as the bear management biologist at Yellowstone.

I'd met Gunther in Mammoth Hot Springs two hours earlier to assist him and his colleague, Jay, a seasonal bear technician, with a whitebark pine cone production survey. Also joining us was Monica, a bubbly high school student interning with the Park Service, who was having the summer of her life.

"Last week I went out with the wolf biologist in Lamar Valley—we saw so many!" she gushed.

Gunther, Monica, and Jay hopped in Gunther's government-issued pickup truck, and I followed in my Tacoma. We parked at a pullout in the shadow of Mount Washburn and hopped out.

"This year's cones are going to be purple and really sappy," Gunther said as we started up a steep, rocky slope in bright sunshine. "Last year's cones will be brown and dried out."

Our plan was to visit two strands of trees, or transects, each consisting of ten whitebark pine trees, and count this year's cones on each.

While grizzlies often get top billing as one of the largest terrestrial carnivores, they are better described as opportunistic omnivores. In the Yellowstone Ecosystem alone, brown bears feed on 276 species—including 175 varieties of plants, 38 kinds of insects, 34 mammals, 7 types of birds, and 4 fish. They'll even chomp on a mouthful of geothermal dirt every now and then for the minerals. Surprisingly, two food sources with the highest caloric value for bears in the Greater Yellowstone Ecosystem are army cutworm moths and whitebark pines seeds.

Army cutworm moths are like M&Ms for bears. The moths feed on the nectar of wildflowers at night and then crawl under rocks in the high alpine during the day to avoid the heat. As with a stream pulsing with salmon, it's not uncommon to see a congregation of grizzlies on these talus slopes in Glacier National Park and Montana's Absaroka Range, turning over rocks and licking up cutworm moths by the thousands.

While whitebark pine seeds don't have the 68 percent fat content of cutworm moths, they come in a close second. Hoping to get the highest amount of caloric return with the least amount of effort, Yellowstone's brown bears let red squirrels do most of the work. While a grizzly or black bear will occasionally scale a tree and knock off entire branches to get to the whitebark pine cones, they'd much rather raid the middens, or food caches, of squirrels who've spent most of the summer stockpiling.

A productive whitebark pine year—often called a "masting event"—is the pine cone equivalent of a super bloom, and it offers many benefits for bears. Whitebark pine seeds not only are an important food source for bears preparing for hibernation but also have been shown to increase cub production and reduce human-bear conflict. The more weight a sow puts on during the summer and fall, the higher likelihood she'll have cubs in the spring. A strong whitebark pine cone year also keeps grizzlies high up in the mountains longer and less likely to encounter humans.

"If it's a good year, they'll spend from August to October eating whitebark," said Gunther. "So, we'll be able to predict bear-related traffic jams based on it because they'll be down lower. Human-caused mortality outside the park will also go up if it's a poor whitebark pine year."

Due to this, a whitebark pine survey is a simple but important tool for wildlife managers and indicates overall ecosystem health.

"Up until 2002, the big worry was white pine blister rust, which is a fungus that came in from Europe on ornamental trees and killed 90 percent of whitebark in Glacier National Park," Gunther explained. "Here, the climate is different, so the blister rust kills branches and the occasional tree, but then the mountain pine beetle came in."

The pine beetle kills by burrowing into the bark of adult trees and eating the cambium.

"We have a beetle like that in Texas," Monica replied. "The Asian Lady Beetle is destroying the oaks."

Historically, the pine beetle wouldn't survive Yellowstone's frigid winters, but with global warming, an outbreak occurred from 2002 to 2009, killing 95 percent of the trees in multiple transects across the park and beyond.

Gunther pointed to a strand of dead trees, looking as if they'd been burned over in a wildland fire. "That big tree on the right would sometimes have two to three hundred cones, and then the beetles got it."

Five more minutes of hiking brought us to our first transect site, where a large capital letter *T* was carved into the bark of each tree.

"Back in the 70s, that's how they marked transects," Gunther said with a chuckle. "We wouldn't do that today."

Gunther, Monica, Jay, and I pulled out our binoculars. The four of us would count all the cones on each transect tree, and if we were all within two to three cones, we'd take the average. Jay, our scribe with the clipboard, would also record any tree with blister rust. Having four people involved in the survey helped with accuracy but also with staying safe in bear country—grizzlies rarely attack groups of three or more people. I was thankful for this added protection: during my second summer working as a paramedic at Old Faithful, Lance Crosby, a nurse at the urgent care clinic in the Lake District, was fatally mauled while hiking alone and without bear spray on the Elephant Back trail.

"Let's go ahead and start with our first tree," said Gunther. "Make sure you're on the right trunk, then systematically work out along the branches."

"Most of the cones are found in the upper part of the tree," added Jay.

We all assumed various positions before a large whitebark pine—standing, kneeling, crouching, lying supine—and brought our binoculars to our eyes, then started counting.

Standing forty feet from the pine tree, I was immediately surprised at how a simple task like counting pine cones could be so challenging. We needed high-powered binos to differentiate this year's cones from last, but that hyperfocus also made it hard to determine which tree trunk we'd landed on. The dense canopy made it easy to count branches (and pine cones) from a neighboring tree, and we'd soon discover there were single trees with multiple trunks.

"I got eight cones," said Gunther, lowering his binoculars.

"I have nine," said Monica.

"Seven," added Jay.

I must've been counting the wrong tree because I counted seventeen cones.

"I initially did that too and got twenty-five," admitted Jay. "So, I moved over, recounted, and got seven."

"Seven, eight, and nine," replied Gunther. "We'll throw out Kevin's and call it eight cones."

"Perfect," said Jay, jotting the number down. "I also didn't see any blister rust."

"Great," said Kerry. "Let's move to our next tree."

I followed Gunther, who, as a biologist in Yellowstone, was following in the footsteps of the legendary biologists John and Frank Craighead.

"They were doing research in Yellowstone, and they had a National Geographic Society special that I watched, and I was really intrigued and said, 'That's what I want to do,'" Gunther recalled as we located the next big *T.* "It was really the Craighead brothers' research that got me interested in bears, and then my first volunteer job got me hooked."

John and Frank Craighead were twin brothers and scientists who'd revolutionized wildlife research and conservation in the 1960s by conducting an ambitious and creative study that spanned thirteen years and three states (Wyoming, Idaho, and Montana) and focused on one species: *Ursus arctos horribilis.*

The twins were born on August 14, 1916, in Chevy Chase, Maryland. From the beginning, they had biology in their blood. Their mother, Carolyn, was a biologist technician, and their father, Frank Sr., worked as an entomologist, studying insects for the US Department of Agriculture.

On weekends, Frank Sr. and Carolyn took the boys and their younger sister, Jean (who would later write dozens of beloved children's books with a naturalist bent, such as *My Side of the Mountain* and the Newberry Award–winning *Julie of the Wolves*), hiking and fishing along the Potomac River, which was an ecological marvel at the time. Deer, bears, and foxes prowled its banks as hawks, owls, and bald eagles soared overhead. John and Frank fell in love with the natural world, learning all the names of the flora and fauna. By high school, they'd developed an interest in wildlife photography and falconry, the practice of hunting using trained birds of prey.

Before heading to college, the brothers traveled west to Jackson Hole, Wyoming. They used their earnings from selling wildlife photos to purchase a Model T station wagon and loaded the car with skis, rock climbing gear, hiking boots, and a few goshawks or peregrines.

Together, John and Frank attended Penn State, where they participated in wrestling and received degrees in science in 1939. With their handsome looks and matching outfits of flannel shirts, rolled at the elbows and tucked neatly into their blue jeans, the twins were inseparable and immensely popular. Following undergrad, the brothers earned their master's and PhDs in wildlife management and ecology at the University of Michigan.

The Craigheads published their first joint article for *National Geographic*, titled "Adventures with Birds of Prey," in 1937, at age twenty-one. Their first book, *Hawks in the Hand*, was published by the Sierra Club just two years later, and in 1940 and '41 they paused their university studies to travel to India for nine months at the invitation of Prince R. S. Dharmakumarsinhji, who shared their love of falconry.

They returned to grad school, but the outbreak of World War II led to the US Navy requesting that they teach soldiers survival skills, leading to the brothers also writing the military manual *How to Survive on Land and Sea.*

Following the war, John accepted a teaching position with the University of Montana, and Frank managed a game reserve outside of Las Vegas, Nevada, but the brothers didn't stay separate for long. Frank married Esther Stevens and purchased fourteen acres in Moose, Wyoming. A few months later, John married Margaret Smith, the daughter of a Grand Teton park ranger, and moved into an identical log cabin next door. Then, in 1959, the National Park Service (NPS) called.

Established in 1872, Yellowstone was originally founded to protect geysers and other geothermic wonders. By 1958, park officials realized that tourists were equally fascinated by wildlife—the wolves, bison, and bears that roamed across America's Serengeti. The park's grizzly population was rapidly declining, however, and bear-related injuries and property damage was on the rise, so the NPS reached out to John and Frank for help: Would the Craigheads like to study Yellowstone's grizzly bears to find out more about the species and why their numbers were declining?

The brothers immediately accepted and moved into the Canyon District of the park, where they converted an old dining hall into their laboratory.

Prior to the Craighead grizzly bear study in Yellowstone, little had been known about brown bears due to the rugged landscape they inhabited, the perceived danger of researching an apex predator, and the assumption that bruins were most active at night. What information people had mainly came from myths, legends, stories, and fairy tales.

At Yellowstone, with the help of Dick Davies and Joel Varney—two engineers who worked in the defense industry—the Craigheads created the first radio-tracking collar for large mammals.

The collar, which housed a small radio transmitter and battery, was composed of fiberglass, resin, and vinyl. It was waterproof, shock proof, and, most importantly, virtually bear proof. The collars worked by emitting pulsing signals the Craigheads quickly learned to interpret. Signals that varied in strength and sound indicated when a bruin was moving. Signal beeps with a constant volume indicated the bear was still, either napping, hunkering down on a carcass, or deceased.

The goal of the Craighead study was to follow the "careers of individual bears from year to year, keeping track of the change in the social organization of the bear community and, in particular, trying to determine the size, composition and grown trends of the population and the diverse factors, both natural and imposed, that influenced it."[1]

---

1. Frank C. Craighead, *Track of the Grizzly*, (San Francisco, CA. Sierra Club Books, 1979), 10.

The first bear the Craigheads collared was a large sow they named Marian on September 21, 1961, in Hayden Valley. Wildlife research would never be the same. "We had acquired the capability of arranging a meeting between bear and man at a place of our choosing," Frank later wrote in his classic book, *Track of the Grizzly*.[2]

After sedating Marian with a drug called Sucostrin that morning in 1961, they assigned her the scientific number of 40 and quickly measured and weighed her. They obtained blood and tissue samples, placed an ear tag to help identify her in the field, and then secured the radio-tracking collar around her neck which immediately began emitting a pulsing signal.

"Hearing this sound meant that we were monitoring the first free-roaming grizzly sow to be tracked by radio," Frank exclaimed, triumphantly.[3]

Marian, the sow, became a muse for Frank and John. The brothers spent eight years following her, gleaning unique insights into the brown bear world and hoping "to learn as much as possible about the grizzly's social organization, feeding habits, seasonal movements, extent of individual ranges, breeding age and frequency of breeding, size of litters, and causes of mortality," they wrote.[4] "We wanted to try to calculate longevity, observe hibernation and prehibernation behavior, and record data on man–bear confrontations."

The ability to track free-roaming animals in the wild forever changed wildlife ecology and conservation. Along with radio-tracking collars, the Craigheads also pioneered wildlife immobilization techniques. To do this, the twins employed a culvert trap—a large metal cylinder mounted on a trailer with a drop-door and baited trigger to trap the bear.

Once the bruin was inside, Frank and John poked a pole syringe through the cage to sedate it. The work was dangerous and the stakes

---

2. Craighead, 17.
3. Craighead, 14.
4. Craighead, 21.

high: if the Craigheads erred on the high side of the sedative dose, they risked sending the bear into respiratory failure and cardiac arrest. If they administered an insufficient dose, the bear could wake up early and be angry and attack. John and Frank quickly learned that Sucostrin's efficacy varied greatly based on a bear's age, stress level, weight, and health.

In other words, it got a little "sporty" in Hayden Valley more than once. Back then, pepper spray hadn't yet been considered for bear defense, so the Craigheads carried a .357 Magnum but never used it, and no one from their team was ever mauled or killed. They usually resorted to climbing a tree or sprinting to their maroon station wagon to escape angry ursids. Once, while National Geographic was filming *Grizzly*, a documentary about the brothers, Bear 36—a big boar named Ivan—woke up early and rammed their car before crawling onto the windshield in a semi-conscious state as the brothers sped away.

In addition to the collared bears, other bruins were in the area.

"In one small timbered bedding area, we jumped thirteen grizzlies," Frank wrote of one encounter.[5] "The forest seemed to erupt with bears."

The brothers also had no qualms about crawling into bear dens to study the architecture and sleeping bruins. They discovered that grizzlies coated the floor with a carpet of moss and pine boughs, they preferred to build their dens on north-facing slopes, and they rarely used the same den twice.

One time, the Craigheads locked themselves in a culvert trap and baited the bears *outside* the enclosure so they could take close-up photos of grizzlies while staying safe. On another occasion, they observed a grizzly following a scent trail 8.5 miles directly to a carcass, revealing a brown bear's incredible sense of smell.

The Craigheads tracked more than two hundred and sixty grizzlies over twelve years and bestowed on them such nicknames as Pegleg,

---

5. Craighead, 161.

Scarface, Cutlip, Lover Boy, and, in the case of one particularly stern mama grizzly reminiscent of a soldier, GI.

Upon observing that brown bears primarily foraged on plants, pine nuts, army cutworm moths, bulbs, roots, and tubers, as well as the occasional elk calf or carcass, the brothers introduced the idea that rather than being bloodthirsty, ferocious carnivores, the bears were, like humans, "generalists" who ate a range of foods.

Frank and John's insights into grizzly bears' nutrition, reproductive cycles, and extensive range were particularly important. By showing that only around 45 percent of sows produced cubs each year ("roughly fifteen animals in the censused population"[6]) and that most females didn't reproduce until the age of five ("though many may not reproduce until they are eight or nine years old"), they were able to illustrate how the deaths of even a few females could have population-wide effects. They also discovered that brown bear cubs have a very high mortality rate.

The Craigheads also illustrated the large and diverse habitats that comprise the grizzly's range as they wander for miles searching for mates and seasonal food sources without regard to federal, state, private, or tribal boundaries. Therefore, management of the species should be ecosystem-based with all the agencies working together toward their survival.

Lastly, and perhaps most importantly, the Craigheads showed the necessity of predators to an ecosystem. They described the bears' beauty in poetic prose, in stark contrast to the negative descriptions by Lewis and Clark. Frank wrote about one late-night encounter near Trout Creek: "The silvery gleam of raindrops on his guard hairs shining jewel-like in the beam of the flashlight."[7]

The working relationship between the Craigheads and the National Park Service forever changed after August 12, 1967, when two women

6. Craighead, 173.
7. Craighead, 22.

at Glacier National Park were fatally mauled by grizzlies in separate incidents on the same night.

The killings were blamed on human-habituated and food-conditioned bears who'd been accessing unsecured goodies and garbage. The common practice of feeding bears at Glacier and Yellowstone was also at fault.

For over eighty years, feeding the black and brown bears who panhandled along the roadside was a great way to snap a souvenir photo, and the bruins routinely feasted at open-air dumps located in the Hayden Valley, Old Faithful, Tower, and West Thumb. The NPS even promoted these attractions, placing signs above the dumps that read "Lunch Counter for Bears" and erecting log bleachers so throngs of tourists could watch the bears as horse-mounted park rangers gave informational chats.

After the women's deaths, Yellowstone's leadership immediately changed their management goals: close the dumps, return all bears to natural food sources, reduce property damage, and lower human-bear conflict and attacks.

John and Frank agreed on the goals but not the implementation. The Park Service wanted to close the dumps immediately, but the Craigheads thought the bears should be weaned off the open-air pits gradually, fearing that an abrupt removal would drive bears into campgrounds, hotels, and employee dorms as they frantically searched for food. This would increase human-bear conflict and the possibility of people being injured or fatally attacked and of bears being shot in defense of life and property.

The Craigheads, famous by then, were not silent in the press about their disagreements, and animosity developed between them and Yellowstone's superintendent, Jack Anderson, and lead biologist, Glen Cole.

The NPS pushed back. Anderson and Cole believed that if they closed the dumps gradually, bears would still pass along foraging strategies to the next generation. They also disagreed with the Craigheads' system of outfitting wild bears with collars and ear tags, which they argued wasn't natural. The Craigheads pushed back, saying their capture-collar program was a small sacrifice that a few animals made to save an entire population.

"The Craighead brothers wanted the garbage dumps to be reopened to save the grizzly, and the Park Service, on the other hand, had the notion that the bears will work it out," said John Varley, a retired chief of research at Yellowstone.[8] "The Park Service said we have faith in the bear. They will find new food sources or old traditional food sources and they'll be fine."

Relations between the NPS and the Craigheads continued to deteriorate. "In the ensuing few years, the climate for independent scientific research in Yellowstone worsened and our work was steadily in various ways impeded, misrepresented and publicly disparaged by park officials because it did now conform to the changed position of management," Frank wrote. "More important, the new policies were very nearly disastrous to the grizzly community."

In the fall of 1970, the NPS decided not to renew John and Frank's research permits. By then, the Craigheads had logged over nine thousand days researching bears in Yellowstone and hiked over sixty-two thousand miles in the name of research.

Sadly, the Craigheads' dire predictions came true in the years that followed the closures of the dumps. As grizzlies invaded campgrounds and public areas to search for food, the sound of gunfire was a nightly occurrence. Some biologists estimate over two hundred and sixty bears died between 1967 and 1972.

In June 1972, a man was fatally mauled when he surprised an adult female grizzly bear who was feeding on his campsite supplies. The NPS responded by creating its own bear monitoring program in 1973. Later that year, the Interagency Grizzly Bear Committee—an interdisciplinary team of scientists and biologists—was created in addition to

---

8. John Varley. Legends of Yellowstone (video). https://www.youtube.com/watch?v=KkCL k0FS8cg

monitoring and researching brown bears in the Greater Yellowstone Ecosystem.

With fewer than two hundred brown bears left in the contiguous United States by 1975, they were labeled an endangered species, and a critical study and recovery plan was conceived.

John and Frank Craighead stayed busy following their departure from Yellowstone—teaching, writing, and continuing to implement the latest technology for scientific research. They modified navy navigation buoys to develop the first animal satellite transmitter, partnered with NASA to use satellites to track wildlife and map habitat, and wrote the National Wild and Scenic River Act, which preserved streams with outstanding natural, cultural, or recreational value in a free-flowing state in perpetuity. In 1998 the Audubon Society listed John and Frank Craighead among the top one hundred figures in conservation of the twentieth century, joining the likes of President Theodore Roosevelt, John Muir, Rachel Carson, Jacques Cousteau, Aldo Leopold, and Henry David Thoreau. After suffering from Parkinson's for years, Frank Craighead died in 2001 at age eighty-five; John lived until one hundred and died in 2016. Their work continues through the Craighead Institute.

By the time we'd reached our second transect of trees—these marked with the removable and more environmentally friendly option, orange tape—I'd gotten a lot more proficient at counting whitebark pine cones. Lying on the forest floor, between abandoned elk day beds, with my binoculars pressed to my eyes, I'd methodically locate the top of the tree and quickly work my way down the branches, without thinking too much.

"I counted twenty-five cones," said Gunther.

"Twenty-three," I replied.

"Twenty-four here," added Jay.

Peter Mangolds

"And I got twenty-five as well," remarked Monica.

"We'll go with twenty-four then," said Jay, scribbling down the data.

Later that afternoon, Kerry and I grabbed lunch at the Wonderland Café in the gateway town of Gardiner. We both ordered a Montana beef burger, topped with white cheddar, basil aioli, lettuce, tomato, and onion, along with sweet potato fries. In between mouthwatering bites, we spoke about the Craighead legacy and how bear management in Yellowstone has changed since the open-air dumps closed.

"The Craigheads were the pioneers of bear research, and when you look at their career and all they did—because bears were just one thing they did—it was amazing. They wrote the *Field Guide to Rocky Mountain Wildflowers*, did a lot of elk and raptor work, and wrote survival manuals for the military during World War II. The legacy with bears was the collaring, and they were also doing the first satellite stuff way back in the 1950s. They were really progressive and pioneers on research, drugs, handling, collaring, and tracking."

"So, who was right—the Craigheads or the NPS?" I asked.

Gunther said they both were. The heavy-handed management right after the dumps closed almost sent the population into a tailspin. "But the NPS was also correct because we didn't teach a whole generation of bears to get into food and garbage, and the population came back," Gunther remarked.

The park has made great strides in the effort to keep bears wild since the 1970s. Today, all the trash cans in the park are bear resistant. Every frontcountry campsite has food storage boxes. Park Service law enforcement rangers conduct nightly patrols through campgrounds, checking for food storage violations. The park switched from a dispersed camping model in the backcountry to designated sites, and each one has either a food storage box or pole on which to hang food. Rangers also patrol the backcountry for food storage violations.

Thanks to these efforts, the current risk of someone getting mauled at Yellowstone is very low.[9] According to the park's website, the likelihood of being attacked by a bear in developed areas, on the boardwalks, or roadside is currently 1 in 59 million visits. The risk to frontcountry campers is 1 in 26 million overnight stays, 1 in 1.7 million to backcountry campers, and only 1 in 232,613 to day hikers. That statistic for day hikers shows a higher risk than for other park users because Gunther estimates only 20 percent carry bear spray, a number he's working hard to raise through messaging and signage.

Despite the success, Gunther knows he can never let up, and human-bear conflict prevention is a "forever thing" he and his team must do. "We're always trying to continually improve our education efforts so tourists do the right thing, and we're continually improving our bear-proofness. We try to make the right thing easy, and we're always trying to think of the next thing that can cause problems and solve it."

---

9. Bear-Inflicted Human Injuries and Fatalities in Yellowstone, https://www.nps.gov/yell /learn/nature/injuries.htm.

Chapter Eight

Ursus arctos on Campus

Each species is a masterpiece, a creation
assembled with extreme care and genius.
—*E. O. Wilson*

Peter Mangolds

The grizzly bear had been under anesthesia for forty-five minutes when she lifted her head, blinked, and looked around.

Crouched three feet away, I stood and took a step back as the bear—a three-hundred and fifty pound sow named Zuri—lowered her head back down.

"I think the drugs are wearing off," I said nervously, bracing myself for an attack.

"Good morning, Zuri," a woman named Ellery Vincent said softly, petting the bruin's head as if she were *Canis lupus familiaris*—a family dog—and not a feared subspecies of brown bear whose scientific name ended with the Latin word for "terrible."

Ellery, a peppy twenty-seven-year-old graduate student, wore a gray T-shirt under her tan Carhartt overalls with a red can of Counter Assault bear spray strategically placed in her right back pocket.

Tony Carnahan, a doctorate student—who looked rather grizzly-like himself with his thick build, rounded back muscles, and long, bushy beard—touched Zuri's eyelid. "I'm testing for a blink response," he explained, "which helps tell us how sedated she is." Zuri blinked slowly and had only a mild reaction when Tony tickled her ear canal, another way to assess responsiveness. "We still have a few minutes."

A veterinarian tech named Jessie, who also worked as an EMT with the local volunteer fire department on the weekends, confirmed that Zuri's vital signs were good before removing the pulse oximeter from the bear's tongue, peeling the blood pressure cuff off her lower leg, and turning off the oxygen cylinder that had been delivering four liters per minute into the bear's nostrils.

As Zuri lifted her head again and gazed around groggily, I took a minute to savor the amazingness of this moment. I'd never stood this close to a grizzly bear or heard the soft sigh of its breath, seen the gentle rise and fall of its chest, or felt its coarse brown fur. The experience was both thrilling and slightly terrifying.

I'd traveled to Pullman, a college town set amid rolling farm fields in southeastern Washington, to visit Washington State University's Bear Center, a one-of-a-kind facility that focuses on research, education, and conservation.

I had first learned about the Bear Center when I happened upon a video of a grizzly walking on a horse treadmill while occasionally being fed bits of hot dogs and apple slices by a wildlife biologist, Charlie Robbins.[1]

The reason the bear was doing laps on the treadmill was a fascinating study into energy use. Animals, like humans, take in oxygen and then use it to oxidize carbohydrates, proteins, and fat. From that oxidation, we produce CO_2, water, and energy that can be used to do work (i.e., climb a mountain or, if you're a grizzly, chase down a wounded moose). The ratios of O_2 consumed, CO_2 produced, and the amount of energy produced are known constants. A scientist can measure either the actual amount of energy that's freed (i.e., heat) or the amount of O_2 consumed or CO_2 produced and multiply those gas values by their known energy constants. "When we put a bear on the treadmill, the bottom of the treadmill below the belt is open," Dr. Robbins explained. "At the top of the chamber is a hole to which is fastened a hose that is connected to a vacuum pump and an oxygen and carbon dioxide analyzer. So, with the use of the pump, we pull regular air in through the bottom of the treadmill, which has 20.98 percent oxygen and 0.04 percent carbon dioxide; the bear breathes in that air, pulls out oxygen for all sorts of metabolic processes, produces CO_2 and energy (work),

---

1. Grizzly on a Treadmill. The Wildlife Society. 2021, https://www.youtube.com/watch?v=OtQWL2MZtYU

and exhales air that might be 19.67 percent oxygen and 1.8 percent carbon dioxide. From those differences, we can calculate exactly how much energy it took the bear to do whatever we were asking."

This research has shown that bears aren't any more energetic than people. "Similarly, they are like humans (and not like wolves) in having a relatively low aerobic capacity, meaning they can't run at top speed for more than a few seconds," Robbins said.

It turns out that bears—like contestants on the History Channel's survival series *Alone*—are always analyzing the energetic cost versus caloric gain of everything they do. "Even big muscular grizzly bears tend not to go straight up mountains but take less demanding paths in steep terrain—which includes using human-built trails," Robbins told me. Sure, they could climb straight up a steep trail, but it wouldn't be wise energetically, so they're much more likely to take a moderate trail with switchbacks—the same sort of trail that we humans would be on. Knowing information like this helps agencies like the National Park Service decide where to put a new trail in bear country or warn hikers of their presence.

Along with energetic studies, the WSU Bear Center was also doing research on nutrition, ecology, behavior, reproduction, learning, memory, and hibernation physiology. When bears hibernate, they don't eat, drink, urinate, or defecate for four to seven months, and yet despite hardly moving, they don't lose any muscle mass. And despite gaining a large amount of weight and maintaining a fairly constant body temperature, their blood sugar remains constant, without high blood pressure or plaque buildup in their blood vessels. Could the secrets of hibernation hold the key to solving human ailments such as diabetes, heart disease, kidney failure, and muscle atrophy? It's a question that keenly interests the biomedical community.

The WSU Bear Center is a single-story, redbrick building, straight out of the 1970s. It is set on a hillside and located about a mile from Martin Stadium, where over thirty thousand people assemble to cheer on the Cougars. I wondered how many of those students were even aware that they shared the WSU campus with, in the words of explorer Meriwether Lewis, "a most tremendous looking animal."

Not that the facility was a secret. When I arrived in Pullman one overcast day in June, I found the parking lot at the Bear Center filled with a lunchtime gathering of people who'd assembled to see the grizzly bears wandering in the fenced-off outdoor areas. The crowd was a mixture of businessmen, dressed down for casual Friday; families with kids; and one woman who sat on a beach chair in the open bed of her pickup truck, pecking away at her computer.

When I asked the woman why she chose to work here, she replied, "I love grizzlies. It's so calming to be in their presence."

The bears exuded a strange kind of kinetic gravitational pull that captivated everyone, me included. One large grizzly with silver-tipped fur prowled around the overgrown grass of the outdoor exercise yard, which was scattered with large tractor tires, downed trees, and exercise balls, all of which provided enrichment activities for the bears. The entire area was surrounded by tall fencing and a sign that read, "DANGER: WILD ANIMALS: Do Not Approach."

Three other bears roamed the outdoor runs that were attached to the buildings and led to their inside dens. The outdoor runs also had tractor tires in them and a large feeding tube filled with water.

At 1:30 p.m., the side door opened and Robbins appeared. In his blue jeans, golf shirt, and ball cap, he could easily pass for an ordinary guy in his sixties rather than someone who spent most of his waking hours with some of the largest terrestrial predators on earth.

"Great to meet you. You can call me Charlie." Robbins greeted me with a firm handshake. "Come in, we're about to start our afternoon lecture."

Charlie first established the WSU Bear Center in 1986, when federal and state biologists expressed a desire to use captive bears in their studies. "The US Fish and Wildlife Service and other agencies had tried to work with brown bears in zoos but had not had any success. Certainly, at that time, nearly forty years ago, zoos had little interest in research," he told me as we walked inside. "So I was asked if I would be interested in establishing a captive facility where a broad range of biological research could be done that would support conservation of wild bears."

Prior to working with bears, Robbins had studied captive and wild deer, elk, moose, and mountain goats, but he dreamed of working with a large predator. "I've always felt that we could learn so much more by working with both captive and wild rather than either alone," he told me as we walked down a narrow hallway lined with framed articles about bear research that had originated at the center, with titles like, "Life in the Fat Lane: Seasonal Regulation of Insulin Sensitivity, Food Intake, and Adipose Biology in Brown Bears" and "Hibernation Induces Widespread Transcriptional Remodeling in Metabolic Tissues of the Grizzly Bear." He especially enjoyed working with bears since "from a research standpoint and with my nutrition interest, I can work with one animal that can be a carnivore, herbivore, frugivore, or omnivore—so just endless possibilities for nutrition studies. Throw in hibernation, and we have a tremendously fascinating animal."

Once Robbins decided to establish a captive bear research program, he searched around campus for a facility that could hold bears. He found an abandoned primate center that was quite dilapidated. "I told the US Fish and Wildlife Service that we needed to start with cubs or yearlings to give us time to learn how to handle them. Of course, when we brought in the first two young bears from Canada, one was out and running through the university dairy herd within twelve hours," he said with a laugh. "So we've been on a steep learning curve ever since and have totally rebuilt the facility."

Today, the WSU Bear Center is home to eleven grizzlies—males, females, and cubs—who were all either born there or brought over by wildlife managers after being orphaned or becoming a nuisance to humans. At the center, bears are trained to take part in humane, approved research. They can open their mouths on command, paw, or nose touch to visual cues, and they willingly present their chests, flanks, or paws for inspection and blood draws, minimizing the need for sedation.

"As I tell my students, bears are really honest and smart animals. When we are cleaning the pens and working with the bears, I tell the students to look at their eyes. You can see what they are thinking and what might happen next by where and how they are looking. They are not sneaky, but they are constantly doing cost-benefit analyses of what they plan to do next," said Robbins.

Robbins has big plans to rebuild and expand the Bear Center to have eighteen pens and three big exercise yards. What would be called the International Bear Center (IBC) would not only increase the space available and the number of bears that could be housed but improve the amount of research that could be conducted. The IBC would continue to house bears that need to be removed from the wild due to human conflicts, but it would also serve local and regional resource agencies by providing a temporary home for injured, orphaned, or sick bears.

"The main challenge is having adequate facilities to keep the bears happy," Robbins said, "and to provide the facilities necessary for high-level scientific research."

His other concern was knowing there are more than fifty problem grizzly bears killed every year for which we could provide a lasting home if we had the necessary facilities.

"Sounds like you've already had a big impact on bear management," I assured him.

"We haven't so much changed bear management as we've helped biologists and administrators understand why different foods or environments are so important to the well-being of both captive and wild bears," Robbins explained. The center also gets involved with projects for the National Parks Service in places like Yellowstone or Katmai when they have a need for information that their biologists can't answer or don't have time.

I followed Robbins into a classroom where a biologist named Justin Teisberg from US Fish and Wildlife Service (USFWS) was giving a lecture titled "Grizzly Bears under Anesthesia" to a crowded, hot room filled with WSU Bear Center staff and a USFWS team from the Cabinet-Yaak Ecosystem in northwest Montana and northern Idaho. The group, which had ventured to Pullman to practice capturing and collaring before their busy summer season working with grizzlies in the field, looked like an REI catalog come to life—everyone wore hiking boots, trekking pants, lightweight puffy vests, and sunglasses dangling around their necks on neoprene straps.

"So we're trying to make sure we are meeting one of our goals during anesthesia," Justin was saying, referencing a slide with an image of a bear's large heart, "which is to match the oxygen demand of the animal with adequate cardiac output and perfusion."

The lecture was in preparation for the following day's lab, when three bears would be sedated and the USFWS team would practice capturing and collaring a grizzly. At the same time, the WSU team would obtain fat biopsies and draw blood for lab cell culture studies—and I would do my best not to look terrified when standing next to a beast with four-inch claws and razor-sharp incisors.

Following Justin's lecture, Dr. Gay Lynn Clyde, director of veterinary medicine, informed us of all the safety precautions in place for the grizzlies. Every bear would be medically assessed on the day of the procedure to ensure that they were healthy and able to be anesthetized safely. Each bear under anesthesia would have a dedicated person to monitor vitals such as temperature, heart rate, and respiration. They would also have a pulse oximeter, a device that measures the amount of oxygen in the blood. If any values were out of range, veterinary staff would be alerted. All monitoring parameters and drugs administered would be documented on an anesthesia record. The bears would also be provided with oxygen and given fluids under the skin for hydration and eye lubrication to prevent corneal drying while under anesthesia. Post-anesthesia, they'd be visually monitored in person and via cameras to ensure their safe recovery.

Lastly, WSU had trained the bears that they'd receive honey if they presented a leg for injection, causing significantly less stress to the animal and allowing the team to administer anesthetics via a hand-delivered intramuscular injection (instead of a dart).

When Dr. Clyde was done, Robbins passed around some handouts, which alerted me that there was more to fear than just claws and teeth. Despite being reminded that there was no rabies vaccine for wildlife or wildlife-domestic hybrids, and warnings that I might be exposed to various zoonotic diseases that I could barely say, much less spell, I nevertheless scribbled my name at the bottom of the page to release WSU from all liability.

The USFWS team and I followed Robbins and Ellery down a long, concrete hallway to meet the bears.

Ellery told me that she'd grown up in Southern California and had loved seeing black bears during family trips to Yosemite. She encountered her first brown bear when she was living in Wyoming's Grand Teton National Park for the summer, after receiving an animal science degree from Fresno State and taking an internship with the American Conservation Experience. "We were about forty miles in the backcountry on packhorses when a juvenile male grizzly ran in front of our group," she recalled. "It was the coolest moment! Grizzly bears are the true emblems of the wilderness."

Ellery was now working on her thesis—"Circadian Rhythms in Gene Expression and How It Relates to Metabolism on a Cellular Level"—and her postgrad goal was either to work in the Cabinet-Yaak Ecosystem on bear management, or stay at WSU to earn her doctorate.

"Every bear that I have wants to weigh a hundred pounds more than they currently weigh," Charlie said as he and Ellery showed us the kitchen and walk-in fridge, where the staff kept medications and donated food that included hot dogs, apples, marshmallows, and the bears' favorite snack—Fruit Loops. "Evolutionarily, the biggest males do most of the breeding, and the fattest females have the highest probability of producing surviving cubs. Combine those two things over four million years, and you get grizzly bears with almost unlimited appetites. They are always looking for that Thanksgiving dinner."

"The three things that drive bears are fear, food, and sex," Ellery said. "Above all and not surprisingly, grizzly bears don't want to be hurt. If they sense a serious threat, they will attack to try to remove the danger. That's why it is often said that if you encounter a grizzly close up, where you can't melt into the distance, the recommendation is to play dead."

"Don't poke the bear. Keep hands, face, and feet away from cages. The bears will bite!" a sign warned us as we entered the corridor where the grizzlies were housed.

There were two bears in each concrete "den," roommates matched according to age and gender. They'd also been administered medications to reduce testosterone, which otherwise caused them to be overly aggressive. Each pen had a window—with heavy metals bars, closely situated, instead of glass and a double-bolted steel door.

"It's like the movie *Groundhog Day*," Ellery joked. "When we release the bears each morning, they always want to test each other to see if the social structure has changed at all."

She gestured to a blonde sow pressing her nose up against the window bars. "This is Willow. She's seven."

Willow shared a pen with Zuri, also seven years old and currently lounging in a large feed tub of water. When she saw us, Zuri took the bone she was gnawing on and placed it onto one of her claws and then appeared to try to flick it at us. We all burst into laughter.

"Zuri is our resident ham," Ellery said with a smile.

Dodge, a male grizzly, typically shared a pen with Adak. Both were also seven years old. Dodge had a blondish eyebrow, a triangle-shaped head, and white claws. Adak could normally be found napping in the corner, but today he was in an isolated pen, eating alone.

"We have to separate Adak because he doesn't protect his food like other grizzlies," Charlie explained, pointing as the bear used his claws to grab a bite of kibble. "And he likes eating his dinner very politely, one bite at a time."

A seventeen-year-old female sow named Kio was napping, holding a bright orange traffic cone to her chest like a stuffed animal. She shared a den with her seventeen-year-old sister Peeka. Both littermates had light ears and other clear similarities.

Nineteen-year-old Luna was a large sow with wild eyes who paced around her solitary den.

Oakley and Cooke, both twenty, were two problem bears. Cooke had been picked up near Cooke City, Montana, the gateway town at Yellowstone's northeast entrance and Oakley came from the Idaho side of the Tetons. She was the smallest grizzly here, weighing about two hundred pounds less than the others, but thanks to her stocky build and feisty personality, she was a dominant bear.

Lastly, John and Frank were both twenty-one-year-old boars who had also been picked up from Yellowstone for becoming too habituated to humans. Fittingly, they were named after the Craighead brothers who had pioneered grizzly bear research in the 1960s.

Although they often took in problem bears, the WSU staff had bottle-raised many of the grizzlies.

"Bottle-raising began about fifteen years after I started the bear program," Robbins said. "When Dr. Lynne Nelson started examining heart function during the active season and comparing it to hibernation—when the bears' heart rate slows to ten to fifteen beats per minute—researchers found that they were unable to distinguish the independent effects of the natural hibernation processes from the possible effect of the drugs they were using to sedate wild-caught bears. So, they decided to start bottle-raising bears since the bottle-raised bears were much more highly malleable and allowed the researchers to simply go in and take ultrasounds."

Robbins said raising grizzly cubs wasn't easy, because they were up every two hours, screaming to be fed. "One of my graduate students that was involved in the process lived in an apartment complex where the rules said, 'no pets.' Because of the screaming, everyone, including the management, soon realized that bear cubs were in an apartment. She managed to keep them by telling the manager they weren't pets. Now the contract says, 'no pets or baby grizzly bears.'"

We all laughed, watching the bruins with wonder and amusement.

"None of the books I've read about wild grizzly bears present their personalities and the thinking processes that determine their actions," Charlie told me. "The great thing about also working with captive bears is that you get to see directly into their lives—what makes them sad, what makes them happy, what they are trying to accomplish on a daily basis, just their big personalities."

"They definitely have cute personalities," Ellery agreed, "but at the end of the day, they're still brown bears and they'll remind you of that."

"As you can imagine, there were a lot of concerns about going in with fully conscious grizzly bears when they hadn't eaten in two or more months," Charlie said, referring to the typical hibernation period around December and January, after feeding stopped in November. "Talk about a knot in your stomach—'Are we going to get killed and eaten?' But being scientists, we always want to learn something new. In we went. What we found even in the older bears was what we called cub reversion."

The researchers found that far from approaching them with aggression, the bears wanted to sit in their laps and suck on their hands or the pacifiers they used when they were cubs. The scientists were able to run tests to determine that the heart anatomy and physiology *do* change: the ventricle thickens, the proteins change to produce a stiffer wall to prevent ballooning due to prolonged filling, heart function changes— and the atria stop beating. "Hard to believe," Robbins said, "but the heart becomes the two ventricles. Just think, the electrical signal starts in the atria but passes over the atria without causing a contraction and then elicits contraction of the ventricles."

As a paramedic who loves cardiology and poring over 12-lead ECGs of the heart, I was astounded. "Bears really do have superpowers!"

I met Robbins at seven o'clock the next morning to feed the bears. The grizzlies were loud, irritable, and obviously hungry. The ferocious sound of paws pounding on metal doors gave me pause as Charlie handed me two white buckets. One was overflowing with frozen apples, and the other was filled with kibbles specifically designed for bears. "Each bear gets one scoop and three Granny Smiths," he told me.

"In the past, the only difference between dog food and bear food was the photo on the packaging," he said, but a WSU study had determined that the ideal diet for a brown bear was 25 percent fat and 20 percent protein. This explains why bears often prioritize eating the fattiest parts of a salmon—the skin, brains, and eggs—and discard the rest of the fish. The fat is necessary to help bears survive during hibernation, while a high-protein canine diet could cause kidney problems in bears.

Knowing this information in the context of grizzly reintroduction efforts in places like Washington's North Cascade Ecosystem is important for wildlife biologists. Historically, bears fed on salmon in the Pacific Northwest to help achieve their ideal diet. But there are worries about whether today's salmon runs are large enough to support them—and if not, is there another fat-rich food they could switch to?

Charlie opened the door to John and Frank's den, a fifteen-by-fifteen-foot enclosure. The bruins were held in their outside run as I hurried inside, but the sound of them banging on the door was deafening. I was certain they'd burst in at any moment and maul me.

I quickly tossed three apples and a scoop of kibble in one corner for John, did the same for Frank, then dashed out. The moment I double-bolted the door closed, Charlie raised the vertical door that led outside, and the bears charged in, diving into their morning breakfast.

Charlie and I continued down to the next pen, to feed Oakley and Cooke, and so on down the line. We fed all the bears that morning except for Luna, Zuri, and Willow, the three who were scheduled to be sedated. Just as with humans about to go under anesthesia, it is important to keep their stomachs empty, lest they become sick while under and aspirate their vomit.

By the time the morning feeding was done, the rest of the WSU and USFWS teams had assembled, and it was time to sedate the bears. I followed Ellery and Tony to the outdoor runs, past the horse treadmill where the bears walked for energetic studies, to a section of fencing with a small square opening about the size of a shoebox at the bottom. All the bears—even those born in the wild—had been trained to extend a paw through the opening in exchange for a sweet reward.

"Up tall, Luna!" Ellery said, raising a bottle of honey water with a plastic straw sticking out.

As Luna hurried over, she looked massive yet positively human standing tall on two legs.

Ellery said, "Good bear!" and rewarded her with a spray of honey water in her mouth.

Ellery lowered the bottle, and Luna sat.

"There you go, Luna!" she said, rewarding her with another squirt. "Good bear!"

Now it was time for the medication. "Leg, Luna. Leg!" Ellery called.

Luna scooched forward, extending her paw through the fence opening.

"Good leg! Good bear, Luna!"

Working at her feet, Tony quickly found a vein and injected the medication to put Luna to sleep.

Ellery and Tony repeated the process with Zuri and Willow. Then the team of observers assembled in the office, watching the bears on security cameras and waiting for the drugs to kick in. Within minutes, Zuri wandered over to her den and passed out. Willow and Luna went to sleep in the outdoor runs.

"Grizzly coming through!" Tony announced moments later as he and Ellery rolled Willow onto a four-wheeled stretcher and transported her from the outdoor run into her den. They repeated the task with Luna.

With the bears back in their respective dens, the WSU team began monitoring the bears by checking their pulses, assessing the sound of their heart tones, and taking their blood pressure by placing a cuff on a lower back leg and a pulse oximeter on their tongues. Once all three bears were safely sedated and their vital signs confirmed to be within normal limits, the USFWS team took turns estimating the bears' weights and calculating drug dosages, then comparing their estimations to the bears' actual weights and WSU dosages to assess their accuracy.

This was an important gauge to get right, because if they overestimated their calculations, a bear's vital signs could bottom out, causing it to die. However, if they underestimated, the bear might not become fully sedated, and they'd be dealing with a grizzly who woke up on the wrong side of the bed.

Next, the USFWS team practiced putting GPS collars on the bears. Jessie, the vet tech, then taught the team the best way to find a vein on a grizzly bear, how to insert a needle and draw blood for lab work, and how to administer normal saline for rehydration. It reminded me of my paramedic training—but for large predators.

A little while later, Heiko Jansen arrived to obtain fat biopsies and blood from the bears for his lab cell culture studies. His area of research was the physiology and genetics of brown bears, particularly gene expression changes before, during, and after hibernation.

As he used an electric razor to shave off a small patch of Luna's fur, Jansen, a friendly man with glasses and a goatee, told me that he'd originally studied sheep. He'd done his grad work at the University of Illinois on reproductive physiology, particularly seasonal changes and how the brain controls reproduction—but when he took a job at WSU and moved to Pullman, he fell in love with bears and hit the genetic research gold mine. "Bears are so obviously seasonal, but not much was known about them at the time," he said.

Jansen's first experience with brown bears had been while visiting Hallo Bay in Katmai National Park. He'd camped with an electric wire around his tent for protection, often sleeping mere feet from where grizzlies were feeding on sedges, shrubs, clams, and fish.

"By any human standard they're morbidly obese," Heiko observed about the bears' size, "but, for a bear, it's healthy and, moreover, it's required because the bears need to use energy from the fat they store to survive hibernation." It made him curious about leptin, a hormone whose main function is to help regulate the long-term balance between the body's food intake and energy use. "Leptin helps inhibit hunger and regulate energy balance so that your body doesn't trigger a hunger response when it doesn't need energy," he explained. "But in bears, when the leptin decreases, their appetites decrease too. Knowing the neuro mechanisms involved in this could help us solve the obesity problem."

Jansen also noted a connection between the amount of fat a female bear accumulates during hibernation and a successful pregnancy. "The link between the amount of fat during hibernation and the number of cubs that survive is a positive trend," he said. The more fat a sow has, the stronger cubs she can produce. They are born earlier in hibernation and exit the den bigger and more likely to survive."

The area that intrigued Jansen most was how these studies could relate to diabetes research. According to the Centers for Disease Control, diabetes affects over 38 million people in the United States, 96 million are prediabetic and, together, this amounts to over $413 billion in medical costs and lost wages annually.[2] Again, I thought of my experiences working on the ambulance and the unconscious patients I'd responded to with low blood sugar or the ones with bilateral, below-the-knee amputations due to diabetic-induced nerve damage.

"Humans need to take heroic efforts of changing diet, losing weight, and exercising to reverse type 2 diabetes," he explained. "Yet bears switch their insulin sensitivity on and off each year. We are very interested in understanding what the cellular mechanisms are that make that happen. There's some fundamental biological difference we can uncover there, and maybe we can tie that to the insulin story in humans and apply it in some way."

It was unlikely to be a single isolated answer. "It involves the

---

2. Diabetes Fast Facts. https://www.cdc.gov/diabetes/basics/quick-facts.html. April 2023.

environment, endocrine, and metabolic systems," he explained, leaning in to take a small biopsy of fat from the bear's large rump. "But that doesn't mean there is not mechanistic process underpinning it all."

"We are just figuring out where the right places are to look to discover those connections," Joanna Kelley, Jansen's colleague, added.

Kelley, a genome scientist, had grown up terrified of bears, only to become surprisingly entranced by the grizzlies at WSU. "They're so majestic, beautiful, and incredibly intelligent," she said. She told me a story about the former manager of the center. "Just like happy house pets," she said, the bears would grow excited at the sound of his car, knowing his arrival meant feeding time. "They were able to make that connection!"

Kelley also realized that her initial concept of hibernation was all wrong. What is happening in the cells, the body, and the DNA? she wondered. And how could their body weight change so much yearly without ill effects?

"If you and I yo-yoed our weight 50 percent every year, we'd be so sick. These are things other animals don't do, and bears do it adaptively," she marveled.

I nodded. "I've seen it on 911 calls—heart attacks, strokes, and kidney failure."

"That all sparked my love of bears and my interest in studying bear biology," Kelley continued. "With bears, the more you learn, the less you know and the more fascinating they become."

She shared with me her excitement over new technologies, like how they'd been able to adapt Fitbits for bears and implant glucose monitoring systems. "Ten years ago, I would've never thought this was possible. So now we are asking what the important thing is to measure and what technology do we use."

By all accounts, the day was a huge success: The WSU team had biopsied fat and drawn blood samples, and the USFWS crew was now better prepared for its summer work with grizzlies in the wild. As Jansen obtained blood samples from Willow and Dodge, I noticed Zuri lift her head.

"Wakey wakey," Tony said, giving Zuri an intramuscular injection of valium, a sedative to help ease the transition back into an alert state and prevent the tremors that are sometimes a side effect of anesthesia medications. "Let's place her in the recovery position."

As Tony, Ellery, and Jessie slowly rolled Zuri onto her side, Heiko noticed me standing there, gazing at the bear with the utmost wonder. He looked over my shoulder and said, "Physiologically, bears are from Mars. And we want to go to 'Mars' to meet them there and see what we can learn."

While Joseph Campbell explored metaphors as myth and religion in his book *The Inner Reaches of Outer Space*, the scientists at WSU are interested in journeying into the outer reaches of the inner space of grizzlies.

My perspective on grizzly bears was changed by my visit. I'd judged brown bears by their appearance, impressed by their size, strength, and speed, but I'd failed to see their unique personalities and the physiological side to them—the tiny daily miracles occurring internally before, during, and after hibernation.

Later, after returning to my hotel, I couldn't sit still so I threw on my running shoes and went for an evening jog. As I ran, I could feel it—that gravitational pull of the grizzly bears, pulling me closer toward their home on Terre View Drive at WSU. I was seized with a giddy excitement. I felt so thankful for my odyssey into the wild, secret life of brown bears and the research the WSU Bear Center was doing, and its potential to help the human medical community felt huge to me, of Nobel Prize proportion.

I slowed to a walk as I approached the Bear Center. It was early evening by then, just before sunset, and the parking lot was empty. It was only me and the bears, in their outside cages, and the silence of the golden hour—an intimate moment between man and beast, when eyes meet and each sees some recognition of themselves in the other.

Chapter Nine

Trouble with Taquka'aq

If you're going to be a bear, be a grizzly.

—*Mahatma Gandhi*

Jennifer Smith

At 2:00 a.m. on May 12, 2023, a firework rose into the sky above the Alaskan Native village of Akhiok and detonated with a *bang*, briefly bathing a tiny collection of homes, fishing boats, and four-wheelers in red.

But this was no celebration. Rather, a skinny subadult brown bear, post-hibernation and hungry, was taking a nocturnal tour of the community, potentially threatening people, property, and pets. When the bear heard the noise, he raced out of the village, disappearing into the night.

Akhiok is an Alutiiq village at the southern end of Kodiak Island. The Alutiiq/Sugpiaq are one of eight Alaska Native peoples who have inhabited the south-central coast of Alaska for over seventy-five hundred years. Today, six Native communities are scattered across the Kodiak Archipelago, accessible only by aircraft or boats. Akhiok, the most remote community, has around twenty-five households with sixty-three residents who practice a subsistence lifestyle of fishing and hunting.

I'd arrived in Akhiok the previous day with Shannon Finnegan, the bear biologist with whom Meaghan and I had searched for dropped GPS collars, and Amy Peterson, a community affairs liaison with the Alaska Native corporation Koniag, to find the community reeling from a bear attack. According to residents, earlier that morning, a bear, *taquka'aq* in Alutiiq, had wandered into the village and eaten a family dog who had been leashed outside.

"We heard a commotion outside and two dogs barking," said Teyo, the dog's owner, when he picked us up at the Akhiok International Airport, which consisted of a single gravel runway and wood rain shelter. "And then there was only one dog barking."

Teyo had raced outside after hearing the commotion and saw a big boar wandering away. "He was the biggest bear I've ever seen," Teyo said. "Easily ten feet."

"Wonder if it was that big boar we spotted from the plane two bays over," Peterson mused.

The flight to Akhiok, which began by us climbing atop the wing to enter the small Piper Cherokee single-engine plane, was rainy and bumpy but also heartbreakingly beautiful: majestic fjords cut deep into jagged, snowcapped mountains, and the herring spawn turned the North Pacific waters a dazzling shade of turquoise.

Since my last visit to Kodiak, Finnegan had visited me and Meaghan in Jackson Hole to connect with some local bear biologists and attend a Ryan Bingham country music concert. The following morning, as Finnegan and I loaded my truck for an early spring hike to Taggart Lake in Grand Teton National Park, I tossed my pack into the backseat, only to have the bear spray in the side pocket detonate, spraying us both in the eyes.

The pain was intense and immediate.

"Get water! Fast!" Finnegan cried, stumbling out of my truck.

Blinded by the spray, I fumbled onto the porch and into the kitchen to fill two Nalgene bottles with water. We spent the next thirty minutes irrigating our burning, bloodshot eyes.

"I can't believe you bear-sprayed the bear biologist!" Meaghan exclaimed when she arrived home later that day. Turned out, the trigger guard was missing on the spray, causing it to activate when I tossed it into the backseat.

"Lesson learned," I remarked during the flight to Akhiok. "Make sure the trigger guard is attached and the bear spray is not in the passenger compartment of any car or airplane."

Finnegan was a great sport and told me not to worry. "Every biologist I know has gotten sprayed at some point," she said in her Irish accent, laughing. "Occupational hazard."

Our festive mood left when we arrived in Akhiok and learned of the dog mauling hours earlier. There were reports that the men of Akhiok had applied to Alaska Department of Fish and Game (ADF&G) to obtain a permit for a subsistence hunt. This would allow them to shoot the offending bear if they spotted it on the Kodiak National Wildlife Refuge, which bordered near the village.

Finnegan couldn't believe our misfortune. "What a day for the bear experts to show up and give a 'bears are great' presentation."

The community school is the largest building in Akhiok and, according to John Stark, who teaches there with his wife, Christina, is the heart of the community.

"Welcome," John said warmly, holding open the door as we loaded in our gear. "Thanks for making the trip down!"

A large blue building with a yellow roof, the school had fifteen students of varying ages and grade levels, who learned in classrooms, a spacious library, and a gym that also doubled as a cafeteria. This year Akhiok will have one high school graduate, Angel Aluska, seventeen, who planned on attending college in Sitka and wanted to travel abroad.

How do the Starks teach fifteen pupils with grade levels ranging from kindergarten through high school?

"Lots of individual instruction," John said with a laugh.

The students were exceedingly polite and respectful. They looked us in the eye as we met them and had names like Leilani, Speridon, Kiersten, Virginia, Jazmine, Juliana, Joseph, and Jarius.

Finnegan and Peterson begin their program by drawing a large grizzly on the dry-erase board and wrote, "How do we feel about bears?"

"You tell me, and I'll write it down," Peterson said, holding a marker.

"And then we're going to ask you tomorrow after our presentations and see if anything's changed," added Finnegan.

"I am mad at them today," said a third-grade girl.

"We don't like it when they attack," added a boy in the sixth grade.

"They have teeth!" said another.

"They killed our dog."

It was clear the bear attack had rattled the children, leaving them hurt and sacred.

Finnegan and Peterson acknowledged their feelings and apologized for the loss of the dog, and then Finnegan gave a presentation titled

"The Bear Necessities: Evolution, Ecology, and Physiology of Brown Bears."

"The first bear-like animal appeared around twenty million years ago," Finnegan began, widening her eyes for effect. "*Ursavus,* the dawn bear, was a small, doglike bear found in North America, Europe, and Asia," she told the students.

Brown bears evolved in Asia around two hundred and fifty thousand years ago and likely wandered over the Bering land bridge about one hundred thousand years ago and into the contiguous United States around fifteen thousand years ago. "They are now found in Europe, North America, and Asia and can live almost anywhere—forests, high alpine, tundra, deserts, and grasslands," said Finnegan. "The biggest bear to ever walk the land in North America was the giant, short-faced bear, *Arctodus.* It weighed over two thousand pounds, could look a six-foot man in the eye, and stood over ten feet tall. As for the Kodiak bear, it wandered across the ice sheets from the Alaska and Kenai Peninsulas twelve thousand years ago and has been genetically isolated ever since. They are the largest brown bears on earth."

When one student asked why Kodiak bears were so big, Finnegan briefly told them about phenotypic plasticity, the ability of an animal to keep getting bigger and bigger—like a child's soft-foam magic grow capsule—when there is abundant food.

"That's why coastal brown bears in Alaska are so much larger than grizzlies found in interior North America at places like Yellowstone and Grand Teton," I added. "Plus, their larger size and rounder shape help keep them warm."

Finnegan's lecture was fascinating, but the students listened with half attention, distracted from the lesson by their more immediate concerns.

Following class, Finnegan, Peterson, and I explored the community of two dozen single-story gray homes with blue roofs that had side shacks for smoking salmon and small fishing boats and four-wheelers parked out front. Beyond the homes, an elaborate Russian Orthodox church sat beside a small graveyard. The first Russians to settle on Kodiak in 1784 had been fur traders and trappers, bringing their guns, germs, and religion with them.

Since there were no hotels, stores, or restaurants in Akhiok, the community school would also serve as our lodging for the night. After a dinner of hot dogs and chili—school lunch leftovers—Finnegan, Peterson, and I each chose a classroom to sleep in. After pushing desks aside, I inflated my air mattress and crawled into bed, feeling like I was in some Alaskan version of *Night at the Museum*.

Peter Mangolds

I awoke to the news that there'd been another bear incident—this one involving the juvenile grizzly in the village—and predicted another rough day for our presentation.

In Akhiok, every school day began in the gym with the Pledge of Allegiance and a land acknowledgment: "We would like to acknowledge that the land we live, work, learn, and gather on is the original homeland of the Alutiiq/Sugpiag people," we all said together.

Next, we sat cross-legged around a traditional Alutiiq oil lamp for morning circle. After three minutes of John and Christina Stark leading us in relaxing breathing exercises, we went around the room to share how we felt that morning and what we were thankful for.

"I'm tired," one student began, holding the tiny speaking stone that was passed around. "I woke up at 2:00 a.m. and I wish bears would stop coming around the village so I don't hear people on the radio in the middle of the night going, 'There's a bear by your house.'"

Everyone in Akhiok stayed in contact via VHF Radio, channel 79, which was broadcast loudly in everyone's home, 24-7.

A kindergartner was next. "The bear will never be near me because the men are going to hunt for that bear," he began. "At least, that's what my mom said."

"I woke up at 2:00 a.m.," said the next student before passing the speaking stone.

"I woke up at 2:00 a.m. too!" said a fourth grader.

"It was another eventful morning," added a woman named Marcella Amodo-White, mother to several of the kids. "I've been up since 2:00 a.m., and my ears are on high alert because I had our windows cracked open, so I was just listening to everything, including my husband snoring."

Marcella grew up in Akhiok, but her husband, Glyndaril White Jr., had been raised as a Raiders football fan in Compton, California. They'd first met online seventeen years earlier. Glyn had immediately fallen in love with both Marcella and the community of Akhiok and had moved to the village to live a subsistence life. The couple and their five children were all featured on National Geographic's show *Life Below Zero: First Alaskans* in a segment depicting Glyn learning hunting, fishing, and the ways of the Native elders from Mavin Angot, Marcella's uncle.

"I was the only one up in the middle of the night because of the radio," the next student said. "I heard the bear was right outside Trevor's house, and then my dad woke up and said, 'I think the bear's been shot.'"

"It was a firework," said the student next to him.

"It was a *loud* firework," added another.

The morning circle reminded me of the critical incident stress debriefings (CISDs) we held at the fire station after a difficult 911 call. Having two bears in their village over the last two nights was traumatic for the kids, and they were struggling to make sense of it.

When John, the teacher, received the speaking stone, he noted the irony of our arrival. "It feels like the grizzlies in the area have somehow gotten wind that we have some bear researchers here," he began, and the students smiled for the first time that day. "I don't know how the bears found out the rumor, but I'm very pleased at the timing."

The Starks had taught in other remote places, like the Yukon, before Akhiok, and knew the importance of using any opportunity to have a guest lecturer to break up the monotony of small town school life.

But we all realized that it would be a tough pivot for Finnegan and Peterson's presentation on bears. So, when John asked if I could begin by giving the kids a lecture on being an author, I eagerly agreed.

Calling up a PowerPoint I'd given before to elementary schools in the Jackson Hole area, I taught the children about the writing process, about creating engaging characters and writing a story that is "uniquely familiar" by using one the seven basic plots: the quest, rags to riches, voyage and return, comedy, tragedy, rebirth, and—perhaps most fitting for our visit to Akhiok—overcoming the monster.

"Dracula and Grendel are some examples of monsters," I said. "Can you think of any others?"

"Bears!"

We all laughed, and I tried to turn it into a teachable moment, reminding the kids that many times the "monster" is someone who, at their core, is hurt, vulnerable, or afraid. "Look at Darth Vader when his mask comes off or the Wizard of Oz when the curtain is pulled away,"

I said. "If bears are monsters, what do they need? In what ways are they vulnerable, and how can we help them?"

Next, Finnegan spoke about some of her research and presented a lecture titled "Staying Safe in Kodiak's Bear Country." Maybe we could help the kids process their scary experience and educate them about having a bear-smart community—where garbage and food are secured—and staying safe in grizzly country.

She began by offering some reassurance. "Grizzlies are potentially dangerous," she said, "but in the last ninety years, there's only been one death from a Kodiak bear attack."

From there, she explained that bears are opportunistic omnivores eager to devour grasses, roots, salmon, berries, and the occasional whale that washes up onshore. And she explained the signs of stress in a bear, the importance of making noise when you hike and of steering clear of sows with cubs or kill sites.

"It sounds like the bear last night was a subadult," Finnegan said, mentioning that bears between the ages of two and a half and five years old were the juvenile delinquents of the bruin world.[1] They'd been kicked out by mom, but were not yet sexually mature, and were now forced to forage on their own and find a home range. Along with being vulnerable to starvation, they were also at risk of being attacked by larger adult males.

As Finnegan finished, Peterson appeared, dressed in an authentic-looking grizzly bear costume. The children laughed as she pranced around the room, stealing miniature apples and candy off their desks to teach a lesson about securing food.

After a short break we convened in the gym, where each student enjoyed getting to dart Peterson with a nerf gun to immobilize her, at which point Finnegan explained what happened during a grizzly capture-collaring mission.

---

1. Subadult Summer. NPS.gov, https://www.nps.gov/katm/blogs/subadult-summer.htm, July 2017.

The bear spray demonstration was next. Finnegan used inert gas to spray the charging bear—Peterson still in her costume—and we closed the day with Finnegan flying her drone to demonstrate the use of an aerial device and thermal imaging camera for population studies.

"So how do we feel about bears now after learning about them for two days?" Finnegan and Peterson asked, drawing a big grizzly on the dry-erase board a second time. "What have we learned?"

"You can tell their age by taking a tooth," said a seventh grader.

"Do not surprise bears," added the kindergartener.

Their fear and anger from the previous days had noticeably lessened, with scientific facts and best practices for dealing with *Ursus arctos middendorffi* replacing emotion.

"Don't wear headphones while running."

"Females need to eat a lot to give birth."

"Some mommy bears will adopt others."

"It's important to secure food and trash."

The children still had a healthy respect for the risks of living in bear country, but now they had the knowledge and skills to stay safe and had a greater appreciation for the species.

Before we left, Finnegan spoke about career opportunities available for someone interested in bears or wildlife. "You can do academic research. You can work for state or federal agencies, nongovernmental organizations, or nonprofits, in the areas of management, law enforcement, tourism, as a wildlife or hunting guide, or for a Native corporation like Koniag."

The Starks thanked us for flying down to Akhiok, and we hurried to catch our plane before a blinding fog blotted out the airport.

Weeks later, I followed up with Marcella.

"We've had a very action filled couple of nights since you left, with bears hanging out at multiple properties," she said.

She told me that while the men hadn't found the behemoth bruin that had killed the dog, the subsistence hunt had taken down an eight-and-a-half-foot-tall brown bear caught damaging a porch. The bear's hide and skull went to the ADF&G, some residents claimed the meat, and the rest of the carcass had been left in the field for scavengers like eagles and seagulls.

She continued: "I've been hearing that with the longer cold months, it is taking longer for the fish to return. That results in hungrier bears and shorter hibernation periods, and therefore they start becoming more problematic in communities and such in search for food, which makes sense."

Marcella told me the community was pretty good about not letting garbage accumulate in their homes, and they usually take trash straight to the secured dump. She noted the change in bear management today versus when she was younger and the community had followed a one-strike-and-you're-out policy regarding bears.

"Back then, people never waited long enough for them to become a risk to the community and residents," she began. "That practice isn't very much in play today, which in my opinion is fortunate for the bear but unfortunate for us, as it gets in the way of us being able to go out and harvest in fear that our safety is in jeopardy."

I couldn't argue with her. People will never tolerate bears if they feel their families and livelihoods are threatened.

Chapter Ten

Binos and Big Ol' Boars

In a civilized and cultivated country, wild animals only
continue to exist at all when preserved by sportsmen.
The excellent people who protest against all hunting and
consider sportsmen as enemies of wildlife are ignorant of
the fact that in reality, the genuine sportsman is by all odds
the most important factor in keeping the larger and more
valuable wild creatures from total extermination.

—*Theodore Roosevelt*

Jennifer Smith

He was born and bred and grew up salmon fed on Alaska's Kodiak Island. He'd been fascinated with brown bears for as long as he can remember—not only their size and strength but also their fast-twitch muscles and the speed at which they can run, swipe, pounce, and bite. As he grew older, he spent hours watching grizzlies in the field—counting them, analyzing their behavior, calculating their movements, and marveling at their unique personalities. But he wasn't a biologist. He was a big game hunter and master guide, and his name is Sam Rohrer.

I hadn't planned on including hunting when I began my journey into the secret life of grizzlies, but then a surprising thing happened. Numerous biologists and conservationists spoke about the role hunting can play in brown bear management.

"I used to think hunting was so bad and I couldn't understand how it could ever be a good thing, but I've done a complete reversal," one biologist told me. "On Kodiak, hunting is part of conservation, and it's really hard to convey that to the public on why it's so important, but I do think it's one of the reasons bears are tolerated and viewed so well here when, oftentimes, people would not care to have them around."

Even a prominent biologist in the Lower 48 wasn't opposed to hunting. "Biologically, we're good," he told me. "So, people's opposition to it is more about personal values and not really science."

When I raised my concerns, he told me a bear hunt wouldn't be like an elk hunt. "It would be closely monitored, and the numbers would be very low," he continued. "I think with the last delisting proposal, Idaho would've been allotted one bear; Montana eight to ten bears, and Wyoming around twenty."

He also mentioned that if a bear hunt ever returned to the Lower 48, a conservation strategy would be in place. "It'd be an early spring hunt, so the females are in the den—or hanging right outside the den—or a late fall hunt because pregnant females would've already denned up," he explained. "There would also be a female sub quota because a lot of hunters without guides can't tell the difference between a male and female. So we'd have a lower sub quota for females, and as soon as they hit that cap, we'd immediately end the season, even if there was more time to hunt males."

"The people who hate grizzlies and don't want any bears on the landscape need to get over it—they're here to stay," Earl, a fourth-generation Wyomingite, said to me. "And the people who love bears and want grizzlies to keep expanding indefinitely without any management or any problem bears removed—they need to get over it too. What we need to find is the middle ground between these two extremes."

Championing my effort to detail many perspectives on grizzly bears, these—and other—biologists suggested I speak with a master hunting guide in Alaska like Sam Rohrer. "The guides know so much about bears it's just mind-boggling," gushed a biologist, "because they're the ones out there in the field for months at a time."

I wanted to argue, but after doing some research, I discovered it was hunters and sportsmen who began the conservation movement in America, helped establish Kodiak National Wildlife Refuge, and saved the Kodiak brown bear from extirpation.

In 1887 Theodore Roosevelt helped found our nation's first conservation nonprofit, the Boone and Crockett Club, a collection of politicians, businessmen, writers, artists, explorers, scientists, and doctors who shared a love of hunting, fishing, and protecting our nation's land and wildlife.[1] The early members of the Boone and Crockett Club read like a who's who of American conservation, beginning with Roosevelt himself. The twenty-sixth president helped create the US Forest Service, National Park Service, and National Wildlife Refuge System, and he protected tens of millions of acres. Today, the iconic Roosevelt Arch greets tourists at the north entrance of America's first national park, Yellowstone, and there's a national park named after him in North Dakota.

---

1. "B & C Member Spotlight," *Boone and Crockett Club,* https://www.boone-crockett.org /tags/bc-member-spotlight.

George Bird Grinnell, an anthropologist, naturalist, writer, and historian, helped create Glacier National Park and started a conservation group that later became the Audubon Society.

John F. Lacey wrote the Yellowstone Park Protection Plan, later known as the Lacey Act of 1894, which provided legal definitions, laws, and doctrines of what a national park (and the National Park Service) could be. His fish and game laws, later known as the Lacey Act of 1900, prohibited interstate transportation of illegally killed game, ended market hunting, and provided the first steps to recovering a population that is still used to stop poaching today.

Gifford Pinchot, our nation's first forester, served as the chief of the US Forest Service, and Aldo Leopold, the father of wildlife ecology, was a renowned scientist, scholar, and philosopher who wrote the classic *A Sand County Almanac*. More recently, Boone and Crockett members Dr. Lee Merriam Talbot authored the Endangered Species Act of 1973, and Russell Train, a tax lawyer turned conservationist, worked as undersecretary at the Department of the Interior, helped create the Environmental Protection Agency, and served as president of the World Wildlife Fund.

With the help of these sportsmen—and many others over the years—initiatives like the Duck Stamp and Pittman-Robertson Act have allocated billions of dollars for wildlife and land conservation. In 1934, the Duck Stamp was created to protect wetlands and waterfowl. While bird hunters over the age of sixteen are required to purchase the Duck Stamp annually to use as a hunting license (and admittance to wildlife refuges), the collectible can be bought by anyone wishing to support conservation. To date, the Duck Stamp has raised $1.1 billion for the acquisition and preservation of six million acres of wilderness.

Since its inception in 1937, the Pittman-Robertson Act (now known as Federal Aid in Wildlife Restoration) has been an excise tax on firearms and ammunition and has allocated over $14 billion toward wildlife projects. Even McNeil River, a wildlife sanctuary that affords brown bears in Alaska their highest protection from hunting, is funded partially by the Pittman-Robertson Act.

While the Alutiiq coexisted peacefully with brown bears for millennia, the Russians had a contentious relationship with grizzlies from the moment they arrived in Kodiak in 1784. When they encountered the archipelago's mild climate, rolling green hills, and rugged mountains, they envisioned a ranching economy based on raising cattle and sheep. As Kodiak bears started killing their livestock and sheep, the bruins were shot, trapped, poisoned, and killed—much like they were in the Lower 48.

"The very numerous and ferocious brown bears on this island must be suppressed," said one rancher.

"Open season should be declared on these carnivorous beasts. Why allow these vicious beasts to run at large?" added another farmer.

"With the elimination of big game, a suitable wool industry could be developed here," one aspiring entrepreneur proposed.[2]

Bears were also blamed by commercial fishermen for low salmon harvests.

In 1937, a wildlife official proposed constructing a large "vermin fence" across Kodiak Island to separate grizzlies from the ranching community. This nine-foot barrier—like the rabbit-proof fence Australia constructed between 1901 and 1907 to keep bunnies from the east out of western farmland—would bisect Kodiak Island to create a bear-free zone.

The fence was built but never erected on Kodiak. The idea was deemed flawed, and the fence was sent to the Kenai Peninsula to assist with a moose research project. Despite the barricade project being scrapped, anti-predator beliefs persisted on Kodiak. Air-raid sirens were used to keep bears away from salmon streams and ranchlands, and dummy grenades were tossed at foraging bruins.

As the killing continued and bear numbers dwindled, sportsmen and hunters verbalized their discontent, which eventually prompted Franklin D. Roosevelt to create the Kodiak National Wildlife Refuge

---

2. Harry B. Dodge, *Kodiak Island and Its Bears*, (Anchorage, AK: Great Northwest Publishing) 2004.

in 1941, setting aside 1.9 million acres "to preserve the natural feeding and breeding ground of the brown bear and other wildlife."

Despite this, the persecution of bears continued, and the anti-bear efforts culminated in a legal but highly secretive operation carried out by the Alaska Department of Fish and Game (ADF&G) to shoot down brown bears with a gun-mounted airplane.

When hunters and sportsmen learned of the plan—involving a semiautomatic rifle mounted on top of a Piper Cub, which shot four inches above the prop and was fired electronically with a button on the control stick, and could be reloaded midflight—and noted a tremendous increase in the number of bears killed, they sent telegrams to the editor of *Outdoor Life* magazine, the preeminent hunting publication at the time.

"The economic as well as aesthetic value of this animal is just as important as any of the renewable resources on the island," one hunter pleaded, "and it should be given the highest possible priority in both the land and wildlife management programs which are developed for the future well-being of Alaska."[3]

The editors of *Outdoor Life* were receptive and ran a cover story in the August 1964 issue. The headline "Kodiak Bear War: Killings with M1 Rifle on Aircraft Arouse Sportsmen" was run with a dramatic photo of a monster-size bear snarling at a tiny airplane. Thanks to the efforts of these hunters, the persecution of the brown bear ended, and residents realized a hunting and bear-viewing industry was more economically viable than ranching.

I thought Kodiak Island might serve as a management model because the archipelago had what was sorely lacking in the contiguous United States—a dialogue between the hunting and science community.

"Sometimes you run into a situation where the biologists don't really respect anyone who doesn't have a biology degree and isn't employed by a government agency," Sam Rohrer told me during a phone interview. "And you get hunters in the field who say, 'That guy's got a

---

3. Tom Kimball, as told to Jim Reardon, "Kodiak Bear War: Killings with M1 Rifle on Aircraft Arouse Sportsmen," *Outdoor Life*, August 1964, 75.

biology degree but sits in an office all day,' and you get this animosity that runs both ways."

"I can see that," I replied.

"But we're fortunate on Kodiak because we don't have that here," Sam continued. "There's a real respect from the sportsmen in the field toward the US Fish and Wildlife and ADF&G staff, and there's a real respect from the staff going back to those that are in the field. They really listen to our observations and see what we're saying."

When Rohrer invited me to join him and his client during a spring bear hunt, I eagerly accepted. While I didn't necessarily need to see a grizzly harvested, I was excited to learn more about hunting and to assist Sam with a field count of bears he was conducting to help the conservation efforts of the Alaska Department of Fish and Game, Kodiak National Wildlife Refuge, and US Fish and Wildlife Service.

Kevin Grange

My Kodiak bear hunting adventure began one rainy afternoon in early May as I found myself flying in Willy Fulton's de Havilland Beaver floatplane and soaring over jagged, snow-covered peaks toward Rohrer Bear Camp, located on the east arm of Uganik Bay. With his white handlebar mustache and easygoing personality, Fulton had a legendary reputation among Alaska bush pilots for being able to fly anywhere in nearly any condition.

"Willy can shovel the bad weather out of the way as he flies," one Kodiak local told me. "And rest assured, if Fulton isn't flying, no one is."

Fulton had over twenty-seven years of flying experience and was expert in dropping off bear viewers, hunters, fishermen, film crews, and snowboarders to Alaska's remote lakes, bays, beaches, rivers, and glaciers. He'd been profiled by PBS and the outdoor clothing company Filson and been interviewed by *Vanity Fair*. He had flown several teams from *National Geographic* magazine and professional snowboarder Travis Rice during his Natural Selection tour.

"Season's kinda late this year," Fulton said, peering down at the snowcapped peaks. "Normally you'd see bear tracks all over up here and empty dens."

It was a windy day with sideways rain, and a somber mood filled the cockpit. My lethargic countenance was likely a result of having one too many drinks with friends at Tony's Bar—Kodiak's biggest navigation hazard—the night before, and Fulton was mourning the death of one of Kodiak's beloved residents. Rubye Blake; her husband, Derek; and their three-month-old daughter were living their dream—spending part of the year in Anchorage and the other half guiding in Kodiak out of their forty-foot fishing boat, the F/V Enchantress. Clients would spend the day hunting and return to the boat at night to enjoy its high-end food, drinks, and amenities.

While Derek and his clients were hunting on Friday, May 5, Rubye fell overboard into the frigid, North Pacific waters of Terror Bay. There was no ladder or step to get back to the boat and her baby. When the hunting team returned that night, they found the baby safe and discovered Rubye's body floating a hundred yards away.

"It's a tragedy, man," Fulton said. "Alaska is one place you can literally die just by walking out your front door."

Fulton spoke from experience. He'd personally known two people fatally mauled at Katmai: filmmaker and activist Timothy Treadwell, who went by the name "Grizzly Man," and his girlfriend, Amie Huguenard, had both been killed at Kaflia Lake in October 2003. Acclaimed director Werner Herzog made the *Grizzly Man* documentary about their story—including interviews with Willy Fulton—and Treadwell's own videos were later released on Animal Planet under the title *The Grizzly Man Diaries* as something of a cautionary tale.

The fact that Treadwell had spent a hundred days each summer around bears for thirteen years shows the large tolerance grizzlies have for humans. But despite over a decade spent documenting bears, Treadwell routinely ignored safety rules, seeming to believe that he had a special, benevolent relationship with the bears and that they looked on him as one of their own.

Treadwell's attitudes showcased his lack of caution. He regularly approached brown bears and turned his back on them to film videos. At Kaflia Lake, he'd chosen a campsite in a bushy area with low visibility, at the crossroads of two very busy bear trails. He never carried any defensive weapons or bear spray. And he'd been fined previously by the National Park Service (NPS) for having food in his tent. All of which ultimately contributed to Treadwell and Huguenard's tragic ends.

Willy Fulton had known and befriended the couple, and sadly, he'd been the one to discover their bodies after the attack. The same bear that killed them had also charged Willy twice: first when Fulton arrived to pick up the couple, forcing him to flee to his seaplane, and again when he'd returned to the site of the mauling with NPS personnel, who shot the charging bruin before it reached them.

As of this time, Timothy Treadwell and Amie Huguenard are the only two victims of fatal bear attacks at Katmai in over a century.

"I've always respected brown bears. I've hunted them in the past, but don't think I would hunt them anymore," Fulton told me. "I also have been very close to being taken out by them, and that goes all the way back to when I was a hunting guide and packer in the Teton Wilderness in Wyoming. We didn't like bears at all; they were a constant threat. They were always stealing elk and we were fighting them off. We did not have a friendly relationship with them."

I asked Willy about the bear who killed Treadwell.

"That bear was mean, man. He was mean to other bears. He always looked at us sideways—just a grumpy old bear, very gaunt—and when he came in there everyone would leave."

Fulton paused, reflecting on that traumatic day.

"That's about as crazy as it gets," he continued, "when you see a bear eating people and the same bear tries to kill you. I was not real fond of bears in that phase."

There's an attraction to and siren song with brown bears that you don't see with other apex predators. Few oceangoers spot a great white shark and have a desire to swim up to it. In Africa, tourists don't step off the safari jeep to snap a selfie with a lion. And yet, at McNeil River State Game Sanctuary, the guides must tell the guests not to pet the bears, and when I worked as a paramedic at Yellowstone, Yosemite, and Grand Teton, I watched hundreds of people approach grizzlies and black bears. I've felt this attraction, too, when I met Honey Bump at Doug and Lynne Seus' house and visited the WSU Bear Center and felt a desire to scratch Adak's snout.

As we approached the Rohrer Bear Camp—a collection of small cabins on a majestic point overlooking Mush Bay—Fulton said he liked Treadwell and will always have fond memories. "He was weird. But he made it thirteen years out there. It's not easy going out there camping by yourself. You feel naked. I've walked around a lot out there and you just feel a lot more vulnerable."

The Rohrer Bear Camp sits on a remote spit of land on the northwest side of Kodiak Island. The camp was previously owned by the Kodiak's first official hunting guide, Charles Madsen. In 1928 Madsen set up a series of silk tents with heating stoves, gas lamps, folding cots, and "only the best brands of canned meats, fruits, and vegetables," and he promised "a trip you will remember with delight."[4] Over the years, cabins replaced the tents, and Charles passed his business (and the camp) down to his sons, Roy and Alf.

Legendary hunting guide Dick Rohrer (Sam's father) purchased the business in the 1970s, founding the Rohrer Bear Camp with his wife Sue and expanding the hunting business to include sportfishing, wildlife viewing, and photography trips.

Sam's fondest memories growing up occurred at camp, where he was mentored by the older guides and learned all about hunting and fishing. Rohrer received his assistant guide's license when he turned eighteen and his registered guide license at twenty-one; he became a master guide at thirty-three. Rohrer has been leading hunts since 1998 and today serves as president of the Alaska Professional Hunters Association. In 2015 he bought the camp from his father and now runs it with his wife and four kids.

"Welcome to camp," Sam said, extending a hand as I hopped off Fulton's floatplane. "Let's give you a quick tour."

Kodiak's bear management plan is viewed as a success story by wildlife managers across the globe, and the bear hunt is highly regulated and founded on fair-chase principles. There are two hunting seasons each year, from April 1 to May 15 and October 25 to November 30. Around two hundred bears are harvested each year from a population of thirty-five hundred. The bag limit is one bear per person every four

---

4. Dodge, 126.

years. Sows with cubs may not be shot, and while a hunter could take a male or female, boars are preferred. All nonresident hunters must have a guide or go with a second-degree relative who is a resident. All hunters must check in with ADF&G before and after their hunt. Hunters may not use bait, hunting dogs, artificial light, or night vision devices; they are forbidden to hunt on the day they fly in (lest they spot a boar from the air) or to shoot on, from, or across any roadway. And all hides and skulls must be sealed and tagged by ADF&G before they leave the island.

Hunts for an out-of-state resident typically cost between $20,000 and $40,000, and according to the ADF&G website, there are pros and cons to each hunting season. Spring hides have longer hair but are more likely to be rubbed (or have bald spots) than fall hides. In the fall, bears usually have more lustrous, uniform coats, and parallel open seasons for deer, elk, goat, and small game offer opportunities for combination hunts, but hunting time is reduced due to waning daylight.

Surprisingly, bears are the one big-game animal in Alaska where hunters don't have to harvest the meat. Based on the people I spoke with, the meat tastes gamey with a stale salmon finish. Even the Alutiiq, Kodiak's original residents, essentially stopped eating bear meat once deer, elk, and bison were introduced.

"So, the meat goes to waste?" I asked.

Sam shook his head no. "The carcass is picked clean in a few days. "You'll see forty eagles on it and occasionally another bear."

During my time at Rohrer Bear Camp, I'd be sharing a cabin with a friendly hunter named Dan who ran a construction business in Tacoma, Washington. I pushed the door open and set my backpack down. The cabin was small but rustic and welcoming. Sleeping bags and hunting gear sat atop bunkbeds lining each wall. Wool socks and camouflage base layers hung drying above a small stove, and bug spray, sunscreen, and hand sanitizer decorated a wood nightstand.

I tossed my backpack on an empty bed, and Sam gave me a quick tour of camp—the outhouse, the cook cabin, and the hides of two giant, freshly harvested brown bears.

"These are from our last hunts. Probably a nine-foot bear and nine and a half. They've been drying, so they're really shrunk and shriveled up," Sam said.

The hides had been salted and hung over a log drying rack. Despite having "shrunken," they were still massive. As I ran my fingers through soft underfur and grizzled, coarse guard hairs, a strange kind of buzzing energy seized me. It was easy to see why Norse berserkers summoned the power of the bear before battle.

"When we talk about rubbed, this is rub," Sam said, pointing to a hairless patch on the hide. "You can see that from a long distance, a little shiny spot. A little short here, common above the tail."

I asked about the process following the kill.

"We'll skin them out in the field, bring them back to camp, and flesh them. It takes a full day to flesh them."

Sam told me fleshing is the process of removing the remaining flesh and fat from the hide and skull.

"Next we salt them, roll them up, dry them, salt them again, and do that a couple cycles because you want to air dry them but must turn the lips and nose," Rohrer continued. "So if you look right here, that would all be the inside of their lip, nose, ears. We turn all that cartilage out. It's a team effort. We all help each other. We take good pride in turning out good bear hides."

Once he's back in town, Rohrer brings the hide and skull to ADF&G. "They'll check its marks, look for tattoos on the lip, lock and tag it. Double-check evidence of sex, make sure it's a male. They'll take a hair and flesh sample off its skull, they'll pull a tooth. They tag the hide and skull, and all that gets cataloged with the hunter's information, the area it was taken; the tooth goes off and gets aged. They catalog all that data and build a database. Every bear that gets killed on Kodiak goes through that office and they check them."

My hands reached for paws and claws the size and sharpness of steak knives. "That could do some damage."

"Oh my goodness, yes," Sam replied. "Seeing the carcasses and the musculature in the arms, the claws, and the articulation, and how they can move them, and then watching them in the wild doing stuff, it's amazing. They can swat, they can grab stuff with those claws. They can't pick up with them, but they can grab stuff and just hook their claw in and move stuff to them. They can really articulate them well. It's so amazing how fast they can move and switch directions and just . . . they impress me."

I followed Rohrer over to three bear skulls sitting on a stone wall.

"One of the things you're looking for is the sagittal crest, and you can get a rough idea on age based on how deep that is," Sam said, handing me a skull that appeared seventeen to eighteen inches in length. "It's kind of elusive, but as the bear ages, that sagittal crest gets higher and deeper so we can measure that. Some guys use that as a ballpark."

Sam handed me another skull that still had a few red flakes of flesh hiding in the eye sockets.

"This is an exceptional one. Look at that crest. See how much deeper it has started to get? So this is an old bear, an exceptionally old one. Look at the receding gum lines, rotten teeth, and front teeth missing. I'd suspect this bear is eighteen to twenty years old."

The goal of a hunt was to shoot the big ol' boars—the ones with rotten, broken teeth who'd already contributed to the gene pool and were most apt to prey on young cubs.

After the tour, I wandered up to the cook cabin where I flipped through years of Bear Camp family photo albums and met Jeanne Shepherd. She's lived in Mush Bay for forty years and worked at the Bear Camp as the chef. Shepherd made tasty meals for a camouflaged crowd who liked to know what was on their plate. In other words, she wouldn't be serving any paninis for lunch, and if she were to present a pastry crust filled with savory eggs, melted cheese, and sausage for breakfast, it would be referred to as hunter's breakfast pie instead of quiche.

"I see hunting as a management program, which I didn't initially when I first moved out here," Shepherd said. "Years ago, if you saw one bear, you'd talk about it all week long, but now you can go up the river in September and see sixty-five bears. You see ten sows with three cubs each. If the bear hunters want to take bears, they want a big bear. And when you take the larger bear out of the picture, now it's the next larger-sized boar trying to kill off the cubs, and maybe now the sow can defend it. She can kill him or run him off, and the cubs survive, so those cubs are growing up and having cubs, and now the population has exploded since I've been here, and that's a fact."

Around 10:00 p.m. I heard the hum of a boat motor, and the hunters and guides appeared. Along with Dan, the other hunter at camp that week was a lawyer from Lincoln, Nebraska, they called Big Tony. As for the guides, they were all bearded and burly and dressed for the elements. If they looked as if they'd stepped off the pages of a Filson catalog, it was because they had. When the popular outfitter wanted to feature true Alaska outdoorsmen, they called the Rohrer Bear Camp guides and bush pilot Willy Fulton.

We sat down at the dining table. Dick Rohrer led us in grace, thanking God for our family, friends, good health, the land, and animals, and wishing us safety on the hunt, and then we dug into Jeanne's delicious spaghetti dish, topped with venison meatballs in a mushroom gravy sauce.

One of the guides, Brett, took out his cell phone. "Sam, you need to see this," he said. "We stumbled upon a bear carcass this afternoon. It was missing its head and paws."

"Was the hide still on?" asked Sam.

"Yes," Brett replied, handing over his phone. "I've got photos and the latitude-longitude coordinates of where we found it."

Rohrer's brow creased with concern as he saw the photos: a headless grizzly and four bloody stumps where paws should be.

"Definitely could be poaching," he said. "I'll send these photos to Officer Boyle with Alaska Wildlife Troopers and call him tomorrow."

I'd been at camp for only a few hours and already it was shaping up to be quite an adventure.

I couldn't wait to get into the field the following day and see some bears.

Fifteen hours later, I sat in a camp chair at the mouth of a scenic estuary, glassing the surrounding shoreline and mountainsides with my binoculars for big ol' boars.

"Key in on alder patches," suggested Brett, who had previously worked as a Hoist Operator and Mechanic on a U.S. Coast Guard search-and-rescue helicopter for over fifteen years. "Like great white sharks, they like to patrol the dark areas."

"The only time you'll see them in the open this time of year is when they're chasing a sow," added Rohrer, glassing the hills.

When Brett described hunting for a Kodiak bear as "99 percent boredom and 1 percent terror," he wasn't wrong. The process to hunt a grizzly consisted of scanning the hillsides for ten to twelve hours each day until you spotted the bear you wanted. Once you had the bruin in your sight, you'd begin to stalk him, trying not to lose the bear in the process. As you closed on the grizzly, the slightest sound or shift in the wind direction could send him fleeing, and according to Rohrer, the bears always had a predetermined escape route and could just vaporize or disappear into an alder patch, never to be seen again. Eventually, you'd want to get within a hundred yards to take the shot, using a .375 H&H Magnum Rifle with 270 or 300 grain soft-point bullets.

"On the pursuit, we're trying to outthink them and trying to speculate where they're going to go and where to cut them off and how to outsmart them. And so, the challenge that brings fuels the desire to be out here, and it's also just fun watching them," Sam said. "You see all this activity, and we watch it from afar through our binoculars, so it's not an activity that is being affected by humans. It's different than when we're fishing on the river and we see these bears because we're in close proximity, but when we're hunting them, most of the time we're removed by at least a half a mile away, and so you're watching bears that are interacting with each other without any other influence from humans. That's just super fascinating to watch."

Sam's client, Dan, was adamant about wanting a bear over nine feet, and even a "slippery nine"—or a large bear on the border between eight and nine feet—wouldn't do. The benefit of having a knowledgeable guide like Sam or Brett was that they could determine the size and sex of a bear, based on their shape and morphology, from hundreds of yards away, helping ensure sows didn't get shot. To both Dan and Rohrer's credit, they stuck to their goal—we'd spotted over ten bears wandering in the distant hills that morning and none met the criteria, so we kept glassing.

As we scanned the hillsides in the freezing rain, we were also conducting a field count and survey.

"We had the idea to do an island-wide real-time bear survey with all the hunting guides," Rohrer explained. "Our vision was we'd do it at the exact same time, on a good portion of the island, so in theory, we wouldn't double count any bears. We'd mark down our sightings on a map so when the staff went back through it, they could see where we were sitting at our spotting locations and where we were observing bears. They could look at it and be like, 'Okay, well, here's the total count of bears and the composition of the count,' and then they could see where the bears are over a four-day period, and each day would be stand-alone."

"Sounds great," I replied.

Sam and his team did the study last year as a pilot program and were repeating it this spring. "The goal is really to develop some baseline information. Obviously, we don't see every bear out there, but the hope is if we do this for five to ten years, we can develop a good dataset that could show trends up or down. So that's been a fun thing that's never been done before on large mammals anywhere, let alone in Alaska. We hope we can set a model for other places in the states to copy it."

The integrated population model, or a field count that uses multiple data streams like aerial flights, trail cameras, hair snares, and field counts from biologists or guides, is quickly becoming the standard. Historically, bear surveys were conducted by aerial flights that could be costly and dangerous and weren't always the most reliable. The noise from the plane or helicopter could scare the bears, and aerial surveys focused on open areas such as meadows or rolling tundra because it was easier to see grizzlies, and the forests could get missed.

Brett handed me his clipboard with the survey form, documenting weather, wind direction, visibility, cloud cover, time, distance, direction, age and sex of the bear, and any additional comments. We'd had quite the eventful morning:

0930: East, ½ mile, Subadult

0951: NNE, ½ mile, Young sow

1000: SE, ¼ mile, Sow w/ 1 cub

1045: North, ½ mile, Young boar

1327: NNE, Young boar with sow

Both the poaching incident and the field survey showed me the value of having conservation-minded sportsmen in the field working in tandem with local wildlife agencies.

"We have another sow and cub coming down the hillside over there," Sam said, directing my attention across the estuary. "Just between those two big alder patches."

I pressed my binoculars to my eyes and spotted the pair, foraging on a patch of green vegetation, surrounded by snow. "I see them!"

Brett pointed in the other direction. "We have a courting couple over there."

In the distance, a boar and sow were locked in a firm embrace as they growled and snapped at each other, looking more like a wrestling match than a mating ritual. The courtship of grizzly bears is unique. The sow uses scent to signal she's in estrus (heat), and many times, the male must follow her around for hours or days, habituating her to his presence, before the union is consummated. During this time, the male protects access to the sow, often sparring with other male suitors and sometimes getting run off by a more dominant bruin. Once the female is ready, copulation can take as long as forty-five minutes, with the two bears alternately snapping at each other. During the breeding months of May, June, and July, a sow may have multiple estrus cycles and mates, meaning there can be multiple fathers within one litter. Following mating, the male typically plays no role in parenting, save the rare exception when biologists have witnessed a boar hanging out with a subadult.

Despite the rain and cold temperature, I was having a great time watching bears all day. I almost forgot the real reason we were there. Around 8:00 p.m., heavy fog moved in, so we returned to camp.

Later at our cabin, Dan told me was hunting with a Sako V Synthetic .375 H&H Magnum rifle with a 20-inch barrel and 300-grain copper Barnes TSX bullets.

"I worked in Alaska from 1981 through 2003 and always wanted to go bear hunting," he said. "I read several books on bear hunting and was thrilled by the adventure. I was always too busy working construction during bear season, so it's forty years later, and I'm finally doing a bucket list item."

"How's it compared to other hunts you've been on?" I asked.

Dan said the hunting aspect was different. "I'm not the tree stand type of hunter. I would rather walk and stalk than sit in a stand and wait," he said. "Bear hunting is much closer to tree stand hunting. Bear hunting is a lot of sitting and glassing hillsides looking for big boars."

When I asked what he planned to do if he harvested a bear, Dan's eyes lit up. "I'm getting a full-body mount!"

We were weathered out the following day as brutal wind and rain pummeled the area and unruly whitecaps marched across Mush Bay.

"That's the nice thing about a day as bad as today," Dick Rohrer, Sam's dad, said at breakfast. "You don't even have to ask, 'Should we go out?'"

Dick was mainly retired now, but he still spent lots of time at camp. I often found him throughout the day in the cook cabin glassing the hillsides and thought, *You can take the man out of the hunt, but never the hunt out of man.*

On my last day at camp, we were glassing from a small island with 360-degree views of Mush Bay when I asked Sam how he differentiates boars and sows in the field.

"The first thing we do is look for the length of the neck. A mature boar will have a long neck," he began. "Younger boars and sows do not have long necks. The head looks like it is just stuck right on the end of their shoulders. If there is very little distance between a bear's hump and its head, then is it either a young bear or a sow. Large boars *always* have a long neck. We watch them as they feed, if a boar is feeding on the level or standing parallel to the hill, he will feed out in front of him, at least a foot in front of his front feet. A sow or younger bear will feed between its front feet. When she puts her head down, her chin is almost touching her front claws. But a boar will have at least a one-foot distance between his chin and his front claws."

The second trait Sam assessed was the profile of the grizzly's head.

"Viewed from the side, a boar has a blocky shape to his muzzle," he continued. "The end of his muzzle is squared off, a female's nose will appear more pointy. The boar's forehead is also dish shaped, where his forehead comes down to meet his muzzle is a squared-off end. Put differently, if you laid a straight edge from the tip of a boar's nose to the top of his forehead, you would have a big gap or space where the forehead rolls down past the eyes, and then hits the base of his muzzle. On a female, if you used the straight edge it would touch the entire distance. In other words, her forehead slopes angled down past the eyes to the base of the muzzle, which then tapers down to a more dainty, pointy muzzle. She won't have the blocky, squared-off muzzle of the boar. Continuing on with the head, looking straight on to the face, a boar's ears will be down lower on the side of his head. On a large, mature boar, he could have as much as ten or eleven inches between his ears. The ears also appear smaller when compared to the rest of the head. A female's ears will generally appear closer and stick more on top of its head. The boar's head will appear more wide and square on the side and the female's will appear more rounded on the sides."

The last thing Sam looks for is the width of shoulders and forearms. The shoulders of a larger male will be at least as wide as the front legs and often wider. "If you can see a few inches of shoulder sticking past the front legs, then it is a *big* boar," he said. "A female's front shoulders will never stick past her legs. And a boar maintains thick forearms all the way down to his paws. Whereas a female's forearms substantially thin down at her wrists."

"So first, the length of neck; second, the shape of head; and third, the width of shoulders?" I replied.

Sam nodded.

"And if you are lucky, you can see a few tufts of hair hanging down from his penis sheath when he is walking broadside. This is actually quite evident at a distance, provided the grass isn't too high," added Sam with a chuckle.

As for estimating the size of a bear in the field, Sam once again started with the length of a bear's neck.

"Only a big boar has a long neck," he said. "We will also look at the weight over his hind end, belly, and then, finally, front end. When you see a bear with a bulky rear end, and his belly is hanging low, back at his hind legs, but then it slopes up to a shallower front end and chest, you are looking at a eight-plus-year-old bear, he is nine feet tall but probably not much over nine and a half. As he continues to mature and grow older, that chest deepens, and he starts to grow larger over the front shoulders. From the front view, the shoulders start to grow out wider than the front legs. A bear that has a big heavy hind end, a deep sagging belly that stretches all the way forward to the front legs, and then large wide shoulders, will be a very big bear, he should be nine and a half feet or more and at least ten to twelve years old, probably older."

"Don't many bears have sagging bellies when they're feeding heavily on salmon in the fall?" I asked.

Sam nodded. "Yes, but he won't have that big front end. And if he is young, he might not have the big hind end either. A sagging belly on a fall bear with no length of neck and a narrower front end is either a mature sow or young boar."

Lastly, Sam instructed me to look at the gait. "A big old bear's legs don't move straight forward in a line, rather they swing their hips and shoulders to move their legs forward. They have to move their legs forward in more of an arc, almost swinging the leg to the side and forward. If a bear is walking that way, you better take a closer look!"

I'd just finished snacking on an energy bar when I spotted a hulking black spot moving in the far distance. The beast was visible to my naked eye, which could mean only one thing—a big boar.

"Over there," I said pointing, "about two hundred yards, halfway up the mountainside, past that big rock."

"I see it," said Sam, standing and forgoing his binoculars to reach directly for his spotting scope.

Adrenaline sizzled through my veins as I stood and readied myself for the stalk.

A moment later, Sam lowered the spotting scope. "It's a bear," he said with a laugh. "But not a big boar."

"What then?" I asked.

Sam smiled. "A sow with cubs. They're packed tight together, so I can see how you mistook it for a boar without binoculars."

I had to leave the following morning, but the hunt continued. While I didn't get to experience a stalk or see a bear harvested, my trip was a success: I learned more about hunting and conservation, saw dozens of grizzlies, and observed some early season bear behavior after they've left the den.

Before I hopped back onto Fulton's floatplane, I thanked Sam and Dick Rohrer for inviting me to camp. "I'll definitely be back and would love to bring Meaghan and my parents for a fishing and bear-viewing trip."

Dan got his bear a few days after my departure. It measured nine feet and nine inches, had a twenty-eight-inch skull, and weighed around 900 pounds.

"We'd been glassing ten days for around twelve to thirteen hours a day before we saw the right boar," Dan told me later. "It went from boredom to a hundred miles per hour once the stalk began. You go from sitting in a chair across the bay from the bears to walking through the area with a heavy brush cover full of bears. We pursued the bear up the hill, and as we went up the hill, it started down the hill. We ended up side-hilling and caught up with the bear back down on the tide flats."

"What was he doing?" I asked.

"He was rubbing on a big tree when we first saw him, and we knew he was a big bear. He didn't see us but quit rubbing and started walking toward the bay. I was slightly above him, and he was about twenty yards to my right, seventy yards away. I knelt on one knee and shot him about mid-body, a foot or so behind the front leg. The bullet went through him diagonally and exited right behind his right front leg. The bullet spun the bear, and he headed back up the mountain. He made it about thirty feet before he fell over. I asked my guide Chris if I should take another shot. He said, 'No, I don't see him breathing, just wait.' We took turns keeping one of our guns on the bear as we worked our way down to the bear. He was dead. The bullet had gone through the heart and lungs. We spent almost an hour prepping the bear and taking photos before skinning the bear. We shot the bear around 9:00 p.m. and were back to the camp around 11:00 p.m."

Flying back to Jackson Hole a few days later, I decided I'd never shoot a bear. While I understood the role a highly regulated hunt could play in grizzly management, providing economic benefits and increasing a community's tolerance for bears in some areas, I agreed with James Oliver Curwood, author of *The Grizzly King*, who said, "The greatest thrill of the hunt is not in killing, but in letting live."[5]

---

5. James Oliver Curwood, *The Grizzly King*, (New York: Doubleday, Page & Co., 1916), 2.

Chapter Eleven

The Chonkiest of the Chonky

Those who have packed far up into grizzly country know
that the presence of even one grizzly on the land elevates
the mountains, chills the winds, brightens the stars, darkens
the forest, and quickens the pulse of all who enter it.
—John Murray

Jennifer Smith

With the arrival of October comes Fat Bear Week at Alaska's Katmai National Park and Preserve.

Katmai occupies over 4 million acres on the Alaska Peninsula, 290 miles southwest of Anchorage. The park was created in 1918 to protect the Valley of Ten Thousand Smokes, which resulted from Novarupta, the volcano responsible for the largest eruption of the twentieth century.

The eruption started on June 6, 1912, and lasted sixty hours, spewing out lava and shooting ash over a hundred thousand feet into the air, blotting out the sun. The eruptive volume of ash and pumice was thirty times larger than the 1980 eruption of Mount St. Helens and collapsed the summit of Mount Katmai, creating the Valley of Ten Thousand Smokes, a lunar-esque landscape of thick ash and pumice, sliced with deep river valleys and dotted with fumaroles, openings in the earth where steam vents. Since the eruption had been preceded by a series of earthquakes, giving ample warning, and the blast had occurred in a largely unpopulated area, there were no reported human deaths.

While Katmai's volcanoes and coast are impressive, most people travel there to see bears, especially those feasting on salmon at Brooks Falls. As the fish swim upstream from Naknek Lake toward Lake Brooks to spawn, a small waterfall causes a yearly salmon traffic jam that results in a feast for bruins.

It was Thomas Mangelsen's famous *Catch of the Day* photo, the same one that kick-started my fascination for grizzlies as a boy, that first introduced me and millions of other people to the wonders of Brooks Falls. These days, thanks to a partnership between the National Park Service and explore.org, a multimedia organization that aims to educate and inspire, every June to October people can follow the bears along the river in real time via wildly popular bear cams, numerous solar-powered video cameras placed at strategic locations along the river. From Brooks Falls, the cameras transmit the footage to a direct satellite internet connection and then broadcast to a global audience.

In 2014, park rangers began a contest on Katmai's Facebook page where viewers could vote for their favorite fat bear by clicking the "like" button. But as interest in the contest rapidly grew, with thousands of people following the lives of the bears, explore.org and the National Park Service created a Fat Bear Week web page and hosted online chats about the grizzlies and Katmai's ecosystem.

Fat Bear Week takes place as fall arrives, just before the bears go to den, with twelve Fat Bear contestants matched up in a single-elimination, March Madness–style online bracket to determine which bear will be crowned the chonkiest of the chonky. The bears, of course, know nothing about the contest—or the office betting pools, T-shirts, coffee mugs, and pillows bearing their likeness—and receive no prize.

Except the prize of weight gain, which in the bear world is a top achievement.

"Fat Bear Week is an opportunity for us to consider the challenges bears face to gain enough weight to survive winter. It also allows us to marvel at the success and health and productivity in Katmai's eco-system," said Mike Fitz, a naturalist with explore.org and one of the founders of the contest.

"One of the unique things about watching bears at Brooks River is we get to know them as individuals and identify them and tell them apart by their facial features, coat color, scars, the way they behave, and all that stuff. It's very different than interpreting bears as anonymous bears on the landscape," Fitz told me.

Once I discovered the Katmai bear cams, I noticed a serious decrease in my work productivity. The bear cams were always playing in the background of my computer, and it was simply too exciting not to watch. Which bears would return to Brooks River in June? How would they look? Who will have cubs? What subadults have been kicked out by mom? How do they spend their days along the river? And like some reality TV series, *The Real Housewives of Brooks River*, what dramatic arguments and fights will occur?

As Fat Bear Week 2022 began, I was glued to my computer. Most of the bears at Katmai had been assigned a scientific number by park staff, but many also had names. Would Chunk (Bear 32), a large bruin with two-toned brown fur, narrow eyes, and a nose scar who loved to hang out at the jacuzzi swirl of bubbling water just below the falls, prevail? Or would Grazer (Bear 128), a sow with a potato-shaped body, blonde ears, and a fierce maternal streak, be victorious? Otis (Bear 480) was my favorite. He was a hulking four-time Fat Bear Week champ with a dark coat and floppy right ear who liked to sit alone in his "office" on the left side of the river. Holly and Divot, two scrappy females in their twenties, were also strong contenders. Bears 901 and 747 didn't have official names, but both had gained hundreds of pounds during the summer. This would be Bear 901's first time in "the big dance," and Bear 747 (aka Bear Force One), who had a medium-brown coat with reddish shoulders, won the competition in 2020 and, not coincidentally, shared a name with an airliner.

Fitz says it's no coincidence that the bears' personalities shine through when people can really spend this up-close and personal time with them. "There are a lot of stories and information about the lives of the bears that would be much harder to discern if we weren't looking at them individually. At Katmai, in some cases, they're released from food competition, so they have much more time and energy to express different behaviors," he noted. Contrary to accepted opinion, there are occasions where sows will adopt orphan cubs and raise them as their own or where old boars tag along with two-year-old male bears.

And there's proof that grizzlies have friendships.

"In 2022, Bear 909 and her sister 910 both had cubs, and in late summer, the families just sort of integrated together and we saw them traveling with each other frequently. They played together, fished together, and allowed their cubs to take food from one another.

That kind of close observation can only come from the webcams and knowing the life history of the bears. It helps us tell a story that we wouldn't otherwise be able to tell."

When asked how the grizzlies were matched for the Fat Bear Week competition, Fitz said it took more than hefty girth to be considered; good storytelling was also necessary. "If we just put photos of fat bears up there and didn't interpret those individual bears, I don't think the event would be as successful. It's about learning about those individual bears, how they got fat through the summer, and what challenges they faced that makes it resonate."

Ultimately, Fitz said, "It's a celebration of an ecosystem that is really operating on all cylinders. And you can't ignore the charisma of the bears."

The cameras also allow viewers to watch how bears behave when released from the pressure of having to find food. Like us, once satiated, grizzlies appear to love napping, swimming, and even playing.

As the 2022 Fat Bear Week kicked off, the event entailed international media attention, upsets, and even a cheating scandal. In a semifinal match between Bear 747 and Holly, rangers and explore.org staff noticed over nine thousand votes for the blonde-eared sow just before the ballot box closed. The team was hoping that a celebrity or influencer had tweeted out support online for Holly, but they soon determined the votes were fraudulent and had to be dismissed.

A captcha box was quickly added to ensure the legitimacy of votes, and on the evening of October 11, I was at my computer deliberating whether to vote for Bear 747 or the Cinderella story of Bear 901 in the championship.

Meaghan and I had visited Katmai two weeks before the contest. It had been the trip of a lifetime. The adventure began the moment we stepped onto Willy Fulton's floatplane departing from Kodiak Island.

Jennifer Smith

It was late September, and we'd signed up for a day trip to Brooks Falls with Kodiak Air Service and Kodiak Island Expeditions, run by Willy and his wife, Jennifer Culbertson.

Prior to becoming a guide, Culbertson, a friendly woman in her fifties with dark hair and dark eyes, had spent twenty-four years working as a law enforcement ranger for the National Park Service and the state of Alaska. Her résumé read like a must-see list of Alaska: Denali National Park, Glacier Bay, Togiak National Wildlife Refuge, Wood-Tikchik State Park—the largest state park in America—and Fort Abercrombie State Historical Park on Kodiak Island. As Culbertson had patrolled the parks, she'd realized that what she enjoyed most was educating people about wildlife and helping them navigate the wilderness, so she switched careers and started Kodiak Island Expeditions, with a focus on bear-viewing trips to Brooks Falls and Hallo Bay on the Katmai coast. Culbertson was an expert at interpreting bear behavior, her Instagram account for Kodiak Island Expeditions had thousands of followers who logged on daily to see her stunning images and learn about grizzlies, and every biologist I spoke with highly recommended her.

While Culbertson handled the logistics and guiding, her husband, Willy, owner of Kodiak Air Service, flew the clients.

"Climb on in," Fulton said, directing Meaghan and me to the back of the seaplane, a de Havilland DHC Beaver. The de Havilland was well regarded in the floatplane community for its efficiency with heavy workloads and its ability to fly in any weather, perfect for Alaska.

Three other people crawled in after us: a travel nurse and her boyfriend, plus a woman named Cecilia who was a fan of the bear celebrities at Brooks Falls.

"I'm hoping to see Grazer, Holly, and Otis today!" she said excitedly, as if searching for Jennifer Lawrence, Taylor Swift, and Tom Cruise.

After a safety briefing, Fulton and Culbertson also hopped in and shut the doors. From Lilly Lake, our takeoff spot in town, we flew northwest over Kodiak. With only ninety miles of roadway, most of the island is vast, untouched wilderness. Culbertson told us about the bear century trails we spotted crisscrossing the wilderness.

"These trails are created when bears traverse the same ridgelines, generation after generation," she explained, speaking to us through our headsets. "Over time, the footprints compress the lichens and moss, and the bears continue to step in the same depressions, deepening them even more."

Fulton directed our attention to the other side of the plane. "Couple of mountain goats over there to the right."

Soon, we were zooming over the vast waters of the Shelikof Strait. Katmai and the Aleutian Range appeared like an apparition in the distance, a moody land of volcanoes and glaciers, fire, and ice.

"Hallo Bay is below," said Culbertson, pointing to tidal flats and sedge meadows framed by gorgeous peaks. "We do a lot of our early season bear viewing here, as the bears arrive to mate, dig clams, and graze the sedge meadows."

After Hallo, we flew over the collapsed roof of Novarupta, Mount Katmai's crater lake, splintered glaciers, and the Valley of Ten Thousand

Smokes. It was easy to see why NASA had sent astronauts here to train for the Apollo missions.

Soon, Brooks Camp, a tiny collection of wood cabins huddled near Naknek Lake, appeared below. Fulton brought the plane in for a landing, and we disembarked.

All visits to Brooks Camp began with mandatory bear school.

"You are in grizzly country now," said Kelsey, a park ranger with red hair, freckles, and a friendly smile.

We suspected as much. Just the stroll from the floatplane across the beach to the visitor center had been a maze of gigantic paw prints and towering mounds of colorful, berry-filled scat. The visitor center was a small log cabin adorned with wooden snowshoes on the walls, twenty folding chairs, a chalkboard showing a map of the area, plus bullet points about viewing-platform etiquette.

"This is one of the only places where solitary brown bears tolerate being so close to each other. They do this so they can feast on the all-you-can-catch salmon buffet," Kelsey explained.

In early summer, the grizzlies fished for energetic salmon leaping up the falls and swimming upstream to spawn. As the season progressed, they became more reliant on dead and dying salmon found in the lower river just before Naknek Lake.

"The falls are what makes Brooks Camp world famous," Kelsey continued, "but bears can be anywhere on the trail, along the beach, or in the river."

I could barely contain my excitement. I was about to witness one of the greatest wildlife spectacles on earth and maybe glimpse some of the bears I'd fallen in love with on the bear cam—Otis, Chunk, Grazer, and Holly.

"Honey," Meaghan said, gesturing down to my hand on her knee. "You're cutting off the circulation."

"Sorry," I replied, releasing my enthusiastic grip. "Just excited."

At Brooks, visitors also successfully coexisted alongside the great gathering of bruins due to several highly enforced rules: All food (and even scented items like lip balm and gum) had to be left in a highly secure food cache. Food would be eaten only in Brooks Lodge or in one of the picnic areas, framed with bear-deterring electric wire. We were also advised to keep gear such as backpacks on us always.

"Don't set anything down and walk away," Kelsey warned. "A curious bear might tear these items apart in search of food or fun things to play with."

As for our "little cubs"—children—they should never be allowed to wander away. Toddlers were to be carried in backpacks.

If we encountered a bear on the trail, Kelsey reminded us to stay calm, step aside, and speak softly. "Whatever you do, don't run!" she said, suddenly serious. "That will trigger a predatory response, and you can't outrun a bear."

Before we were dismissed, Kelsey told us the platform etiquette: no smoking, no flash photography, no tripods (monopods only), and most importantly, no loud noises or cheering.

"What type of idiot would cheer?" I asked Meaghan rhetorically as Kelsey dismissed our group.

"That's it?" I asked, standing. "We can just go wander around now?"

I felt as if we'd just been given the keys to *Jurassic Park*.

"We can," Meaghan replied. "At least we have Jennifer and Willy to lead us."

It was a cool, overcast autumn day, and the trees lining the river were a sorbet swirl of orange, maroon, and yellow. Surprisingly, both Brooks Lodge and the Katmai Trading Post had closed the day prior, and the end-of-the-season feel was in the air. The number of bears at Brooks usually peaked in July, when sockeye were leaping up the falls by the hundreds, and the number of tourists did likewise. It wasn't uncommon to have over three-to-five-hundred people a day visit Brooks Camp around then. At times there could be a two-hour wait to get onto a viewing platform and only a thirty-minute time slot allotted.

According to Leslie Skora, an NPS biologist at Katmai, grizzly viewing at Brooks was far different than in the Lower 48 or even a place like Kodiak, where the bears are hunted during the spring and fall.

"The bears here have a much smaller space bubble than you'll find at Yellowstone, the Tetons, or even Denali," she told me prior to my trip. "Because of the abundant salmon resources, they're willing to tolerate each other and people in a lot closer proximity. It's kind of like the New York City of bears, and we're able to have these viewing opportunities that don't happen much else anywhere in the world."

But as Fulton and Culbertson reminded me, from the moment the bears left the den, they lived under a ticking clock. "They have to eat a year's worth of food in six months," Culbertson explained, "and compete with other bears for those food resources."

"They're still dangerous and wild," added Fulton.

During our 1.2-mile hike down a dirt road to the viewing platform at the falls, Culbertson told us how grizzly bears were ecosystem engineers, tilling the soil with their long, excavator-like claws; dispersing seeds via their scat, and scattering salmon carcasses throughout the forest, introducing carbon and nitrogen into the ecosystem.

"They've found salmon DNA in all the trees around there," Culbertson said metaphorically. "The salmon carcasses discarded by bears replenishes the forest and surrounding ecosystem with nitrogen, carbon, phosphorous, and other minerals."[1]

Bears can eat up to thirty or forty fish per day, especially when the bears are in hyperphagia, a state of hunger that can never be satisfied, no matter how much they eat. "They may not eat the whole fish," Culbertson explained. "When there's an abundance of salmon, the bears 'high grade' by only eating the fattiest part of the fish: the brains, skin, and eggs."

After a mile, we reached the elevated walkway leading to the viewing platform at the falls. As we hiked, we saw hulking bruins in various shades of cinnamon wander underneath us on their way to and

---

1. Robert S. Semeniuk, "How Bears Feed Salmon to the Forest," *Natural History,* April 2003.

from the river, or lying on their stomachs, napping with their forepaws forward and hind legs stretched back, a posture known as splooting.

Once we reached the viewing platform, Brooks Falls came into view. Sure enough, there was a behemoth brown bear on the lip of the falls, staring intently down at the water and waiting to catch a leaping fish.

"Do you know who it is?" Meaghan asked.

The large bear had a light brown coat with dark patches around her eyes. "I think it's Grazer," I replied. "Bear 128."

A few grizzlies wandered below the falls, while others patrolled the shore, searching for fish scraps. I spotted one of the Fat Bears, Otis, in his "office" on the left side of the river. With his big belly, he'd wait patiently in his corner—as if in a seated, silent meditation—and then suddenly he'd pin a fish with his mighty paws before grabbing it in his mouth and waddling up the steep riverbank to devour it in the woods.

At first, the bears appeared to be randomly situated along the river, but Culbertson informed us that under the surface, a chess game of hierarchy and dominance was taking place. The biggest, most dominant bears got the best fishing spots at the lip of the falls (or just below in the jacuzzi), and the less-dominant bears were spread out across the cascade and downriver. Dominance was a dynamic based on the bears' age, size, health, and attitude, and it could change yearly, seasonally, or weekly. Typically, the big adult males were most dominant, followed by sows with cubs, then other adult males and females, on down to the hungry and hormonal teenagers: the subadults. These were the curious, scrawny underdogs, struggling to stay alive and remain resilient in a harsh world, and I loved them more for it.

Once Culbertson and Fulton explained the hierarchy, I was entranced to realize how every bear was, in some manner, acting and reacting to all the other bruins in the area.

Since the bears had the same objective—Operation Binge on Fish!—I could see their different personalities based on how they fished. The bear at the top of the falls practiced the stand-and-wait method. Over in his office, Otis preferred to sit and wait. Another bruin searched for fish with his head darting back and forth under-water—a technique known as snorkeling—while another went diving, submerging his whole grizzled body. As for those sweet subadults, they weren't above begging for scraps from more dominant bears or pirating (aka stealing) fish, then dashing off into the woods with their treasure. A few bears also tried the dash-and-grab method, leaping into the river, displacing a lot of water in the shallow section, and hoping to pin a fish with their paws. This tactic burns a lot of energy but, sadly, isn't very effective.

Suddenly, the bear named Grazer, standing at the lip of the falls, caught a leaping fish.

"Yes!" I exclaimed loudly, clapping.

Meaghan was more than embarrassed. "You're *that guy*," she said. "The cheerer."

I was ashamed to have broken viewing-platform etiquette, but it was suspenseful watching the grizzlies fish, and knowing the stakes and the effort they put in, it did make you want to cheer.

After spending an hour on the platform—during which I happily spotted Bear 132, Bear 909, and Holly, along with Otis—we wandered over and sat in the tall grass beside Brooks Lake. A few bears snorkeled and dove around for fish in the distance, and one blond subadult looked like he was just floating on his back, enjoying a moment of rest and the warm sunshine on his face.

Every now and then, we'd hear a noise in the woods behind us and spot an eight-hundred-pound grizzly, ambling down the trail, mere feet from us. And on our hike back to Brooks Lodge later that afternoon, we passed within a few feet of a half-ton, silver-tipped grizzly sleep-ing in the shade. We always tried to keep fifty or one hundred yards between us and the bears, but there were so many of them—and they were so silent—sometimes we just happened upon them, or they upon us. They were still apex predators, and I made no pretenses about their friendliness, but they were far from bloodthirsty killers intent on eating humans. Despite the proximity of thousands of people to the bears at

Brooks, there have been few injuries and no deaths at camp. Frankly, having never been hunted or given access to human food, these bears simply didn't care about us.

Following a picnic lunch—eaten safely behind an electric fence—we returned to the main viewing platform to discover the falls empty. No bears stood on the lip or fished in the jacuzzi below. My heart sank and I recalled the famed conservationist, Aldo Leopold, describing Escudilla in his classic, *A Sand County Almanac.*

Escudilla is a 10,912-foot mountain in Apache County, Arizona, in the eastern part of the state and visible from western New Mexico. Leopold started his US Forest Service career in the Escudilla Wilderness, the site where the last brown bear in Arizona was shot by a predator control agent in 1933 who was given orders to kill bears to prioritize ranching in the area.

"Time built three things in the old mountain," Leopold wrote. "A venerable aspect, a community of minor animals, and plants, and a grizzly. . . . The bureau chief who sent in the trapper was a biologist versed in the architecture of evolution, but he didn't know that spires might be as important as cows. He did not foresee that within two decades, the cow country would become tourist country and, as such, have a greater need of bears than beefsteaks."

Leopold discusses the harmful effect Manifest Destiny had on nature, Native Americans, and animals like wolves, bison, and grizzlies. "It did not occur to us that we, too, were the captains of an invasion too sure of its own righteousness."

With the last bear gone, Leopold writes, "Escudilla still hangs on the horizon but, when you see it, you no longer think of bears. It's only a mountain now."[2]

Gazing out at Brooks Falls without bears, the area had lost its magic for me. I contemplated a future without bears and recalled something Lynne Seus had said to me: "Without wildlife, it's just scenery."

---

2. Aldo Leopold, *A Sand County Almanac,* (New York: Oxford University Press, 1949).

Back home weeks later, I realized I still needed to vote for the winner of Fat Bear Week 2022. While I loved 901's Cinderella story and her first appearance in the big dance, I couldn't deny the size and appeal of Bear 747.

"The Bear-o-plane it is," I said.

A few hours later, just past 7:00 p.m. MST, the results were in. Tallying 68,105 votes over her 56,876, Bear 747 beat Bear 901 by a solid margin to claim the title.

Fat Bear Week 2022 was a resounding success. Over one million votes were cast, and the contest received media attention from the likes of the *Today Show, USA Today, NPR, Wall Street Journal, Washington Post,* and others.

I could feel my perspective on grizzly bears and the natural world changing. Where once I'd seen only an amorphous mass of fur, jaws, and claws, I now observed the coat and claw color, scars, ear and muzzle shapes—and personalities—that made each bear unique.

Leslie Skora, the biologist at Katmai, had suggested I visit the McNeil River State Game Sanctuary, a lesser-known spot at the top of the Alaskan Peninsula. "It's the gold standard and one of the best places on the planet to watch brown bears."

Unlike the hundreds of tourists who came to Brooks Camp, only ten people were allowed to visit McNeil every four days to observe the largest gathering of grizzlies on earth. It wasn't uncommon to see over sixty bears fishing for chum salmon at the falls at the same time.

I logged onto Alaska Department of Fish and Game's webpage for McNeil's permit lottery and applied.

Chapter Twelve

Devils or Deities

There is a bear with glowing eyes deep
in the human unconscious.

—*Carl Jung*

Peter Mangolds

I sat across the table from Larry Van Daele, a former bear biologist with the Alaska Department of Fish and Game, at Henry's Great Alaskan Restaurant on Kodiak Island, and for some unknown reason, we were talking about volcanoes.

"If something is really that big and powerful, are you afraid of it? Or do you worship it?" Van Daele began, taking a sip of pale ale. "Are the people of Hawaii afraid of volcanoes, or do they worship them?"

It was early May, and I was back on the Emerald Island to learn about the spiritual and Indigenous aspect of brown bears.

For thousands of years, Native Americans have coexisted with bears across North America. However, I knew that writing about the traditions, beliefs, and rituals regarding bears in the varying tribes could fill volumes, so I chose to focus on one Indigenous group that had a rich history of living with *Ursus arctos* and still had grizzlies on the landscape, the Alutiiq/Sugpiaq.

My day began with a visit to the Alutiiq Museum on the corner of Mission Road and Kashevaroff Avenue. The Alutiiq are an Alaska Native people and the first settlers on Kodiak. They arrived over 7,500 years ago, paddling over from the southwest peninsula. They hunted and fished, following seasonal food resources much like bears.

"*Aleut* is the Russian word for coastal dweller," Patrick Saltonstall, curator of archaeology, explained as we toured the museum's gallery. "But their traditional self-designator is *Sugpiaq*, meaning 'real person.'"

As we explored the museum, I marveled over exhibits featuring chipped stone tools—drills, knives, and points—and a wooden hand tool with a human face carved on the end.

"This speaks to the Alutiiq belief that there is a humanlike consciousness in all things," Saltonstall explained. "All creatures, objects, places, and elements have a spirit."

A labret was a decorative ornament, inserted below the lip, that men and women wore to display age, family, and social position, and a *qayaq*, or kayak—made of wood, animals skin, and sinew—was a testament to the courage of the Alutiiq people, providing the ability to hunt sea mammals with ground-stone lances and bone-barbed harpoons.

Saltonstall directed my attention to the front of a *qayaq* from 1867. "Look at the artistry and precision," he remarked. "That upturned prow enabled them to glide smoothly through the rough, windy waters surrounding Kodiak."

Next, Saltonstall opened a small drawer beneath the *qayaq* and pulled out a translucent object that had the shape and feel of rigatoni pasta.

"Bear intestine," he said. "The Alutiiq used intestines to make lightweight, water-repellent rain gear."

As we toured the rest of the gallery, I marveled at the ingenuity, creativity, and resilience of the early Alutiiq people.

Following the gallery, I was lucky enough to get a private tour of the temperature-controlled special collections room (normally closed to the public) from Amanda Lancaster, the museum's Curator of Collections.

"This is a traditional oil lamp," she began, showing me a smooth rock that had been sliced in half. "We store the lamp upside down when not in use—to prevent the spirit from escaping."

I loved the idea that spirits animated everything in Alutiiq culture. "It means every person, animal, plant, and stone is sacred," I said.

Over the next half hour, Lancaster showed me all the bear-related artifacts in their collection—dagger-like teeth, shiny skulls, pelts, sinew, and long, white claws.

Next, I spoke with Dehrich Chya, language and living culture manager for the museum. Chyais Sugpiaq from Kodiak. He'd gone to Hawaii for college, then returned in 2018 to work at the museum. Along with his duties at the museum, Chya has been an Alutiiq dancer for most of his life. Alutiiq dances share stories, honor ancestors, thank the animals for supporting them, and pray for a happy, healthy year ahead.

"We've always respected bears," Chya began. "They're smart, and long ago, people believed bears could understand the Alutiiq language when you spoke, so people wouldn't talk about their hunting plans since could understand that."

The Alutiiq hunted brown bears from the moment they arrived on Kodiak. "Bears were a really important resource because there weren't any other large, endemic land animals on Kodiak," Dehrich explained.

In his book, *Kodiak Island and its Bears*, Harry Dodge explains there were strict rules and rituals surrounding the hunt. Only men were allowed to participate, and they were forbidden to have sex the night before. The men took steam baths to clean themselves; they hunted only in clean clothes and wore decorative hats and amulets for good luck. Once they harvested a bear, a hunter never bragged about his kill, and the carcass was handled with the utmost care because the brown bear was considered a human ancestor. The hunter would also place his arm down the grizzly's throat—to demonstrate his lack of fear—and then the skull was placed on a tree, facing east, as a sign of respect.

Later, the village would have a ceremonial feast and return the bear's bones to the kill site as a sign of respect. Unlike today where the meat is left in the field for scavengers, the Alutiiq used all parts of the animal. The bones were turned into tools and arrow points; sinew became thread; the bear's teeth were used in jewelry; the hide made great bedding or a carpet. The guts were used for waterproof clothing, and the meat was consumed, even if it occasionally tasted like stale salmon.

Chyais also edited the book *Uniqkuat: Kodiak Alutiiq Legends*, which shares sixty-two traditional tales from the past one hundred and fifty years, told by Alutiiq elders. "Many of our stories showcase the kind of power bears have and teach valuable lessons," he explained. "We believe people can transform into animals and species can have a human form. If they take off their skin, or beak if it's a bird, they can turn into a human."

One of the stories, "The Woman Who Became a Bear," tells of a woman who transforms into a brown bear by chewing on the snout of a bear (to make it pliable) and stretching the hide over her head. She rips her husband to pieces after she discovers that he abandoned her and their child and cheated on her.

"The White-Faced Bear" tells of an invincible bruin with white feet and a white head who devours a hunter who won't stop killing bears and threatens to make them extinct. In another legend, "The Bear Hunter," a man is "tore up until there is nothing left of him" after he promises never to kill a sow and cub and then kills a mother and her cub the next year.

"What strikes me about the stories is the environmental ethic contained in them," I said to Dehrich. "The legends are entertaining, but they also teach valuable lessons about sustainable harvest and respecting nature."

"Bear with Tails and the Sly Fox" tells about a time when grizzlies had tails. "Bears used to have really long tails, kind of like dogs," Dehrich said. "One day, there was a fox out ice fishing on a lake and the bear shows up and sees the fox, and is like, 'Hey, how'd you catch all these fish?' So, the fox, a tricky animal in our stories, decided to play a prank on the bear. The fox says the bear should use his tail as a fishing lure, so the bear puts his tail in the ice hole, and it freezes over. He didn't catch any fish, and when he stood, his tail popped off."

"Sounds like the story in Greek mythology of how Hera, the wife of Zeus, turns the nymph Callisto into a bear as punishment for infidelity and flings her up into the sky with her long tail and she becomes Ursa Major, the Big Dipper," I said.

Before I left, I bought a water bottle with petroglyphs decorating the side and a coffee mug with the word *Iqualluq* engraved in it, Alutiiq for "place of many fish."

Following my visit, I snapped a few photos of the Holy Resurrection Cathedral, the white Russian Orthodox church with two blue onion-shaped domes, on the corner of Mission Road and Kashevaroff Avenue, and then wandered over to Henry's Great Alaskan Restaurant to meet Van Daele. The renowned eatery is located at the back of a small strip mall with a liquor store, bank, and souvenir shop.

As I strolled through the parking lot, it dawned on me I had no way of recognizing Larry. Just then, a truck pulled in with a license plate that read, "Taquka'aq"—Alutiiq for "bear."

"Hey Larry," I said, hurrying over.

"How'd you guess?" he asked, hopping out and giving me a firm handshake.

Van Daele was a tall, sturdily built guy in his sixties with a salt-and-pepper beard and friendly smile.

"The license plate gave it away," I replied.

Van Daele said when he worked in Dillingham, Alaska, the tribe had two names for him: Larry and Bear. "Yupik Eskimo tribes, like many Native cultures, have two names for people. The first is their given name and the second is their Native name, based on something they've done or resemble. It was an unusual honor for a white boy, especially one that worked for a government agency, to be given a Native name. That's why I put it on my license plate."

Larry had received his master's in wildlife management before later earning his doctorate in natural resource management from the University of Idaho. He'd moved to Anchorage in 1977 and started working for ADF&G. Birds of prey were his primary interest at the time, but after six weeks his supervisors at ADF&G asked if he wanted to move to Kodiak and start a bear project.

"I said sure . . . where's Kodiak?" chuckled Van Daele as we walked into Henry's and grabbed a high-top table in the bar area.

When he started out, Van Daele was putting a tracking collar on the first brown bear he ever spotted in the wild, and as an avid hunter, he had some ulterior motives. "I thought, *They're going to pay me to learn everything I can about bears, and I'm going to get the biggest one out there to put on my wall.*"

"What was your Boone and Crockett score?" I asked, referring to the measuring system trophy hunters used. Van Daele shook his head and said after working with grizzlies for a few months, he could never kill one, a sentiment he's never felt for any other species.

"Bears really captured me and ended up showing me the world because of the experiences I've had because of them," he explained. "So I started out not knowing a lot about bears, and they chose me is the best way to put it."

Growing serious, Van Daele told me about a presentation titled "Devils or Deities?" that he gives to high school students as part of a "Living with Bears" lecture series offered by the Kodiak Brown Bear Trust, a non-profit, non-government organization dedicated to working with the public and private resource managers to promote healthy bear populations, improved understanding of Kodiak bears, and better ways to coexist with them.

"Bears perceive the world differently and because of this and their intelligence, people cannot completely understand what is going on in their heads. This, in turn, can incite fear, especially given their large size and strength," Van Daele said. "So I begin by telling the kids that bears can smell time."

"Please explain," I said.

"Bears have this incredible sense of smell, maybe three to four times better than a bloodhound, or three hundred to four hundred times better than a human," Van Daele began. "If you think about the chemical nature of smelling, chemicals denature at a given rate, so if a bear comes across a trail, it can tell what direction that person was going—left or right—because there's more chemical denaturing on one side of the trail because it was there longer. So, a bear can tell how long ago a person was there, so a bear can smell time."

"Reminds me of the First Nations proverb," I began. "A pine needle drops in the forest; the eagle sees it, the deer hears it, and the bear smells it."

Van Daele also informed me grizzlies could see smells and gave the analogy of how a campfire eddies in one spot as it's going from one direction to another. "Because their sense of smell is so acute, they can essentially see the smell coming off our body when they perceive the world," he explained, "so bears can see smells."

"Can I get you guys something to eat?" our waiter asked.

Larry ordered the Alaska fish and chips and a pint of pale ale.

"I'll do the same," I replied. "And a Guinness, please."

Van Daele continued. "Trying to understand a creature that intelligent, that perceives the world so differently and is so adaptable, it's just fascinating to me. How can we coexist with something like that? It's easy to coexist with a deer, but with a bear, oh man!"

Thirty-five years in the field taught Larry no one can truly be an expert on grizzly bears because they perceive things in such a different realm and the bears often defy expectations. Once, Van Daele darted what he thought was a large sow wandering around with two two-year-olds. But when his partner rolled her over, he exclaimed, "Larry! She's got balls!"

Larry couldn't believe an adult male would be hanging out with a couple of two-year-olds, and he was forced to reconsider existing theories on infanticide.

"The best we could hypothesize was that mom was ready to kick the cubs out and find a new boyfriend, and these youngsters had lived in the vicinity of the adult male," Van Daele said. "So the bear did his thing, the cubs did their thing together, and they fooled the bear biologists."

Another time there were two sows with cubs along a river, and a big boar came in and scattered the families. One mom ran off with her two cubs, but the other sow got separated from her little one, inadvertently crossed the river, and fled. "Once the boar left, the mother with two cubs heard the temporarily orphaned cub bawling and barked it up the tree for safety where it remained until its mother appeared ninety minutes later," Van Daele said, shaking his head with wonder.

Van Daele had worked with bears in varying capacities—on the management side, research side, and political side—but I wanted him to speak about his experiences as leader of the Northern Forum Brown Bear Expert Team and his travels to circumpolar areas like northern Japan, Scandinavia, and Siberia to study Indigenous tribes of Evenks, Samis, Invits, and Alutiiq along with Swedish and Siberian shaman. This was how, surprisingly, we started speaking about volcanoes.

"What really struck me as I worked with those folks is how each one of the tribes in those regions had almost the same kind of relationship with bears, and that relationship on one hand was scared shitless because they could eat them up—and did prehistorically—and the other aspect was these bears are liaisons from the spirit world," Larry explained.

"Cultural diffusion?" I asked. "Or Carl Jung's collective unconscious?"

Arctolatry is the scientific name for bear worship, and Van Daele believed it related to the dichotomy humans have with volcanoes. I was beginning to see where he was going with this. "So tell me, are people in Hawaii afraid of volcanoes or do they worship them?" I asked.

Larry said they do both. "And that's the way it's been with bears because if you look at the way it can walk on hind legs, the way it sits to nurse its young, and if you see a skinned bear—it really looks like a dead human lying there in a lot of ways—so you see a lot of kinship in behavior that people recognize as humanlike behavior, and then you throw on top of that, this animal dies and goes to the grave every fall and then is resurrected out of the grave in the spring. I don't care if you're a Christian or a pagan, that's a pretty powerful symbol."

I agreed.

The symbolic death and rebirth of hibernation was the reason Van Daele believed different circumpolar cultures came to the same conclusion about bears being liaisons between the spirit and human world without communicating with each other.

I took a swig of Guinness and told Larry about my morning exploring the Alutiiq Museum and speaking with Dehrich. "There were lots of stories about humans transforming into bears, bears transforming into people, and the idea that bears could understand our language."

Larry said when he spoke to elders from central and northern Alaska, they don't use the word *bear*—even in legal proceedings—to this day. "You say, *Old Grandfather* or *the Brown One* or some other euphemism to be respectful."

I loved that we were having that conversation amid the flashing TV screens, bar chatter, and beer-battered smell of Henry's Great Alaskan Restaurant.

During his travels, Larry learned about the Ainu of Japan and Scandinavian and Siberian shamans who had rituals regarding the hunt.

"If you kill a bear, you need to do it in the right way because that bear could be your ancestors coming back after dying," Larry explained. "So, you want to be very respectful to that animal if you harvest it, when you harvest it, and how you do it."

Historically, on Japan's northern island of Hokkaido, the Ainu would capture a cub out of the den, raise it as an honored guest in the village for two years, and then, in a very ritualistic and solemn manner, kill the bear and eat it.

"Isn't that disrespecting it?" I asked, plunging a piece of fried fish in tartar sauce.

"From an Ainu perspective, they were showing the bear how wonderful the people were and how wonderful they treated the bear, so when they released it back to the mountain gods, it would say, 'The mountain people are good. Let's give them good crops. Let's not bother them anymore, and let's help them out.'"

Prehistoric Scandinavian cultures would ritualistically harvest a grizzly bear and take out the eyes so the bear's spirit couldn't see them, then they would skin the bruin and place the skull on a tree, facing east. Next, they'd brew a special kind of mead (honey liquor) and pour it through the hole (foramen magnum) in the back of the skull so the mead would flow out through the nose, and hunters would take their cups and hold them up to the skull and drink to get the power of the bear.

"Kind of like the berserkers wearing the hide of a bear," I said, recalling my conversation with Doug and Lynne Seus.

"Know what the word for 'skull' in Finish is? *Skal,*" Larry said, raising his pint. "This later morphed into the word used during a toast. So that's harkening back to the old days when they'd get their mead out of a bear skull."

To prove how reverent circumpolar cultures were to the bear, Van Daele told me about a time he was out in the woods on Kodiak with a Russian biologist, and they happened upon a bear skull.

"Any biologist worth his salt is going to grab that skull, and I said, 'We can get the permits for you to take it home,' and the biologist—this PhD who I'd worked with for years—asked which way was east. I pointed and he took the skull, put it in a tree facing east, said a couple prayers over it, and then said, 'We can go home now.' He felt that strongly about it."

In Alutiiq culture, I'd read they used all parts of the bear because that's the gift the grizzly gives to you, yet today in Alaska, brown bear is the only game animal where you don't have to harvest the meat, and you can't leave the skull on a tree facing east, because you'll get fined if you *don't* bring the skull in from the field to register it with ADF&G so they can check its age and sex.

Finishing off the last of my french fries, I asked Van Daele if we'd lost reverence for the bear today.

"That's where a bottle of Scotch and a campfire comes in," he replied with a laugh, suggesting it would be a long conversation.

When I asked about coexistence, Larry told me about a Kodiak Native Elder who'd taken him under his wing. "What you need to know about bears is very simple," he told Van Daele. "Respect the bear, and the bear will respect you."

I nodded, despite not knowing exactly what the teaching meant.

"That phrase summarizes everything I've learned about bears over the years in one simple sentence from a true Elder," said Van Daele.

As we browsed the dessert menu, Larry told me bears weren't the unpredictable ones—people were—but we couldn't place all the blame on humans. There were intelligent, big-personality bears; some rocket scientist bears; some stupid bears; and also some gangbangers that needed to be removed.

"Even the Ainu people believed that," Van Daele added. "Every now and then you'd get a bear that wouldn't behave, and it was okay to shoot it because it had gotten kicked out by the mountain gods because they didn't respect it either."

Later, as we left Henry's and walked to our cars, Van Daele said he was optimistic that grizzlies and humans could coexist because, if we just gave bears a chance, they can adapt, so we should be able to as well.

"Bears are very powerful, very potentially dangerous, and they could be considered a keystone species in how they fit into the whole ecosystem," he said. "And if we can coexist with something that is, in some ways, equally as powerful as we are—if we don't have our guns—if we can respect and allow them to live with us, then we are allowing a whole suite of other ecological aspects to live."

"Makes sense," I replied.

"By living with bears and letting them coexist with us, we're showing respect to an animal that could be harmful to us, but we are doing it in such a manner that we're taking care of everything that's below bears in the food chain and the whole ecological aspect of that."

Chapter Thirteen

The Ursine Paparazzi

It's not the bear that crosses the road,

it's the road that crosses the forest.

—Anonymous

Peter Mangolds

189

Armed with digital cameras, cell phones, spotting scopes, tripods, and zoom lenses, the ursine paparazzi had been waiting for weeks at pullouts along Pilgrim Flats in Wyoming's Grand Teton National Park for Bear 399 to emerge from hibernation in spring 2023. The crowd arrived in the dark, predawn hours each morning and stayed all day, waiting for the famous bruin to appear. And then, late in the afternoon on May 16, 2023, Bear 399 emerged from the forest, and she wasn't alone. A new cub, with a white natal ring around its neck, trotted a few paces behind her to a chorus of camera clicks, flash photography, and tears.

"I'm so happy that she is still with us," Thomas Mangelsen later told *Mountain Journal*. "She continues to amaze and break records as the most incredible grizzly that has ever lived in our lifetime."[1]

Bear 399's appearance was historic in multiple ways: first, there's only a 9 percent chance a grizzly bear in the Greater Yellowstone Ecosystem lives to age twenty-seven, and few bears that old still reproduce. At twenty-seven, she was now the oldest documented mother in the Greater Yellowstone Ecosystem, and this was her eighteenth cub spread out over eight litters.

Since 399's historic walkabout in November 2021, she emerged from hibernation in spring 2022 and immediately headed south, leaving the boundaries of Grand Teton National Park possibly to search for the type of human food she'd received just before denning. But a wonderful thing had happened: while she slept during winter of 2021–22, wildlife officials, concerned citizens, and local nonprofits like Bear Wise Jackson Hole and JH Bear Solutions had begun addressing Teton County's trash problem through education and outreach efforts at the neighborhood level. Consequently, without any human-bear conflicts

---

1. Thomas Mangelsen, as quoted in Todd Wilkinson, "Famous Jackson Hole Grizzly 399 Wows Again, But Now What?" *Mountain Journal*, May 18, 2023.

in the local community, 399 returned to Grand Teton National Park later that spring, spent a few more weeks with her four cubs, and then kicked them out as aggressively as she once defended them. Bear 399 stayed mostly hidden for the remainder of the summer, but she was spotted nuzzling up to a large boar named Brutus just before she went to the den in the fall.

One week after 399's miraculous reappearance with her cub, who was named Spirit by bear watchers, I parked my truck at a pullout just past the Pilgrim Creek bridge. Dozens of cars lined the road, and I could tell from the large number of people it meant one thing—a bear. Grabbing my binoculars and camera, I hopped out and joined the crowd.

"Is it 399?" I asked excitedly.

"There are two bears in the woods," a man replied. "It definitely could be!"

"I can't wait to see Spirit!" cried another woman.

"Three-ninety-nine is amazing," added a man wearing a shirt that said: *399: The Original Mamma Bear.* "She is risen!"

Traffic was backed up on the park road, and at both ends of the bear jam, members of Grand Teton's volunteer wildlife brigade, clad in yellow fluorescent vests and each with a can of bear spray on their belt, directed cars and people.

"Keep moving, please!" said a man to a Honda that had stopped in the middle of the road. "Both tires over the white line," he called to a woman parking a sprinter van.

A third volunteer stood amid the crowd of people, struggling to keep them back. "Please stay next to the roadway," she said to some college student inching away from the road. "At least a hundred yards away from the bear."

I hurried over, peering into the forest and hoping to catch a glimpse of America's most famous bear and her spring cub.

Peter Mangolds

Twenty-four hours earlier, I'd spent the day with Justin Schwabedissen, Grand Teton's bear biologist, to learn more about roadside bears and the park's renowned wildlife brigade program.

I met Schwabedissen at eight o'clock at the Craig and Thomas Discovery and Visitor Center—CTDVC in NPS speak—in Moose, Wyoming, and hopped into his government-issued pickup.

"Lots of exciting bear activity this spring," Schwabedissen said with a handshake.

Schwabedissen grew up in Idaho Falls, studied wildlife resources at the University of Idaho, and received his master's from Utah State University. His thesis was on mule deer migrations, but he quickly grew to love another transboundary species—*Ursus arctos.*

Schwabedissen first worked in the Tetons as a Wildlife Brigade intern in 2011 and after ten seasons as a wildlife technician was hired as a biologist, one of only six full-time grizzly bear biologists in the National Park Service. Despite the stresses of large carnivore management in a national park with four million tourists, Schwabedissen loves his job.

"Every day is different. There's never a dull moment, and that's what keeps the job so much fun," he said as we drove north on Highway 191.

"To be working with grizzly bears and wildlife every day in the Teton Range, I couldn't ask for anything different. There is always something to learn, new ways to approach a visitor, better ways to articulate something, or just a place I probably passed a thousand times but didn't look at it the way I did one day."

Grand Teton is unique and offers some of the best grizzly bear viewing outside of Alaska.

"For many visitors, it may be their only opportunity in life to see a grizzly bear in a natural habitat," Schwabedissen said.

"Do they forage close to roadways to avoid adult males who want to eat the cubs?" I asked as we passed Snake River Overlook, site of Ansel Adams's famous black-and-white image of the Tetons. "Is it for the human shield effect? A food source? To reduce competition?"

"We currently don't have a lot of evidence to support it one way or the other," Schwabedissen said. "We do know that the more dominant bears—typically the males—are able to access the best habitats across the ecosystem, possibly forcing less dominant bears like sows with cubs into marginal habitat along the road and developed areas of the park."

While the male bears and wolves feed on winter-kill ungulate carcasses like elk, deer, and bison in the early spring, sows with cubs and subadults migrate to the meadows in Pilgrim and Pacific Creeks to feast on the "green up"—vegetation like biscuit root, a member of the parsley family with a tuber on the end that looks like a primitive carrot.

After the open-pit dumps were closed in 1970, the first habituated bears feeding on a natural food source in the Greater Yellowstone Ecosystem appeared in Yellowstone National Park in the early 1980s when hikers spotted them digging roots and foraging along the Mount Washburn trail.

"People were so freaked out, they just assumed these bears were troublemakers, so initially we moved some, hazed a lot of them, and then eventually we realized that these bears are just doing bear things and the only difference was they were doing it with a hundred people standing there," said Kerry Gunther, Yellowstone's bear biologist.

"The bears were learning to live with us, so I said let's try managing the people so we can allow the bear to access the meadow, and the public just loved it. It was the best thing, so we tried managing the people instead of the bears. Along the way, I questioned if it was a good decision, but it was because the public loves it, and the roadside bears almost never get into any trouble."

As grizzlies recolonized Grand Teton in the late 1990s and early 2000s, grizzly sightings along park roadways became more common. In 2006, grizzly 399 emerged with three cubs-of-the-year. As more visitors came to the park specifically to see bears and individual animals like 399, new challenges emerged for park staff. People were approaching grizzlies, trampling fragile habitats, and displacing the bears, and the traffic jams through Pilgrim and Pacific Creeks became routine. Plus, there was always the real possibility someone would get hit by a car or mauled.

Initially, law enforcement rangers worked the bear jams, but it quickly became apparent they were too busy with their primary duties to handle the animal jams full-time as well.

"We want people to have the opportunity to see these bears, we just have to ensure it's happening in a safe manner," Schwabedissen said as we passed the entrance station in Moran and drove through Pacific Creek.

"There's always a balance we're trying to strike between providing bear viewing opportunities for the public but doing so in a way that protects the resource and doesn't harm the very animals visitors come to see."

In 2007 Grand Teton started a wildlife brigade pilot program that consisted of one permanent bear management specialist, one seasonal wildlife manager, and one volunteer. In the years since, as 399 and other sows continue to raise their families by the road—and thousands of people travel to Grand Teton specifically to see the bruins—the wildlife brigade has grown to one permanent bear biologist (Schwabedissen), two seasonal wildlife management rangers, and over thirty volunteers who contribute over thirteen thousand hours in the park each year. While the wildlife brigade's focus is on the bears, they also handle elk, pronghorn, great gray owl, and bison jams.

During spring and summer, the wildlife brigade patrols the park from 6:00 a.m. to 10:00 p.m., seven days a week, facilitating safe wildlife viewing at wildlife jams (areas of traffic congestion resulting from visitors stopping to view animals) and patrolling lakeshores and campgrounds for unsecured food and garbage, along with providing educational programs at kiosks in the pullouts.

The volunteer commitment varies: many "volis" live locally in Teton County and work six-hour shifts. Other volunteers commit to working thirty-two hours a week in the park on eight-hour shifts and are given an RV spot in the park for the summer.

The volunteers come from all walks of life. Many are retirees after having very successful careers, but they all share one similarity. "They just want to give back, and they have a deep passion for bears and the park," explained Schwabedissen. "They love the Tetons, wildlife, and we just can't thank them enough."

The wildlife brigade's season is typically between April and October, when bears and other wildlife are most visible, and the majority of the funding comes through the Grand Teton National Park Foundation.

"If the animal is attracting a crowd of people, we'll usually show up and help make sure that animal can move unimpeded and that people are maintaining an appropriate distance of a hundred yards for a bear or wolf, and twenty-five yards for all other wildlife," Schwabedissen added.

Once accepted into the program, volunteers complete a compre-hensive training program on bear safety and ecology, wildlife jam management, and the messaging necessary to handle visitors who sometimes don't like to be told no and occasionally push the envelope. In the event a visitor gets too unruly or isn't compliant, the wildlife brigade can always call a law enforcement ranger. The primary duties of the wildlife brigade are to enforce the rules: no stopping in the roadway, park in a pullout or alongside the road with both tires over the white line, and don't feed or approach wildlife.

During their shifts, wildlife brigade volunteers drive white SUVs with yellow emergency lights and wear ball caps, sunglasses, and reflective vests for safety. They carry bear spray, a radio, and rangefinder, which measures the distance from the bear so they can be consistent in enforcing the hundred-yard rule. Much of their job is getting people to understand that they're in bear country and there are risks, but visitors can do a lot to mitigate those risks to stay safe.

The wildlife brigade volunteers have different reasons why they love their job.

"Probably the most exciting thing is working with grizzlies and giving our guests the opportunity to get some beautiful camera shots," said Larry Muir, who's been with the brigade for ten years.[2]

"Visitors are just so thrilled," added Laurie Wofford. "Maybe the animals want to cross the road and here's a line of people, so the biggest challenge is to nicely move them back and make sure that we get cooperation without disappointing or without making people upset."[3]

Wofford said it's the staff she works with, the people who support the brigade like the Grand Teton National Park Foundation, and the visitors who are so thrilled to see the animals that make her job fun. "They thank us for keeping them safe and allowing them to watch wildlife. Ethical viewing of wildlife is our main job. I think the wildlife brigade is essential in protecting the wildlife as well as the visitors."

As we passed Oxbow Bend, a crescent-shaped section of the Snake River overlooking the Tetons, Schwabedissen said there are numerous bears using the roadway corridors this spring—399 and Spirit; Bear 610 and her three yearlings; and 926 with her couple of two-year-olds.

2. Larry Muir, as quoted in Grand Teton National Park Foundation video, "*A Day with Grand Teton's Wildlife Brigade,*" February 2015, https://vimeo.com/120022805.
3. Laurie Wofford, as quoted in Grand Teton National Park Foundation video, "*A Day with Grand Teton's Wildlife Brigade,*" February 2015, https://vimeo.com/120022805.

Biologists in the Greater Yellowstone Ecosystem don't bestow names like Felicia, Bruno, and Spirit on the bears. Rather, the bruins are named by the millions of fans who follow them in person and online.

"We want people to appreciate bears as a species and remember there's a vibrant bear population across this ecosystem," Schwabedissen said. "Yes, it's great to see individuals and learn their story over time, but we really want people to focus on the species in general."

At Jackson Lake Junction, Schwabedissen hung a left and we pulled over at a bear jam that was just breaking up.

"We just saw 610 and her three cubs," said one man. "It was amazing."

"Makes all the travel to get here worthwhile!" added a woman.

A man in his sixties approached our truck. "Any idea where 399 is?"

"Probably off foraging somewhere," Justin replied. "But I'm sure we'll see her again."

With the bears gone, the two wildlife brigade volunteers were focused on moving traffic, their hands waving cars through.

"Need anything?" Schwabedissen asked them as we slowly drove by.

"Nothing," they replied. "We're in the Tetons watching bears. What could be better?"

As we turned around and headed north, Schwabedissen explained another bear management strategy: seasonal management closures. As an example, Willow Flats, the wide expanse of land in front of Jackson Lake Lodge, is closed from May 15 to July 15 every year for elk calving season and the grizzlies who like to feed on them, Schwabedissen explained.

"Bear 610 knows when elk calving is occurring and where to find calves bedded down as mom forages elsewhere; she'll start a grid search looking for those little guys."

As we passed Pilgrim Flats, cones lined the road, designating a no-stopping zone. Once upon a time, this strip of asphalt would've been lined with cars, but the park has identified places where 399 and other bears usually cross the road and has sectioned them off. "Visitors are able to park outside the no-stopping zones and sometimes walk back to see a bear, but they must stay at the roadway and a hundred yards away from any grizzly," Schwabedissen explained.

Justin also pointed to a dirt lane cutting perpendicular to Highway 191—Pilgrim Creek Road.

"This is another temporary mitigation closure. This avoids the bears being boxed in on two sides," Schwabedissen explained. "If we get a line of parked cars along the highway and then get another line of vehicles down Pilgrim Creek Road and the bear is in the middle, this disrupts natural bear behavior, so we temporarily close the side road and keep everyone along the highway as a group to allow the bear to move and forage unimpeded."

From Pilgrim Creek, we continued north and turned left into the Colter Bay area, consisting of a campground, concessionaire employee dorms, a marina, general store, cafeteria, ranger station, and access to Jackson Lake, a shimmering mirror reflecting the Tetons.

"In most areas of the park, bears have the right of way and are provided space to move across the landscape as they wish. We may close areas, such as around known carcasses, to minimize human disturbance, provide for visitor safety, and protect natural interactions between wild animals," Schwabedissen said as we drove down to the lakeshore. "However, in developed areas like Colter Bay, we prioritize visitor services. Due to the large number of visitors in developed areas and the high potential for human-bear conflicts, there is no tolerance for bears lingering or foraging in these areas. We will allow bears to move through developed areas to get from point A to point B, but otherwise we actively discourage bears out of the developed footprint."

To deter the bears from spending much time in developed areas, the Park Service uses a variety of hazing techniques. Schwabedissen defines hazing as a temporary negative stimulus to temporarily change behavior. Soft hazing tactics might be clapping or a brief chirp of a siren to softly push a bear walking down the road off the asphalt for its own safety. When soft hazing techniques are not enough, park staff may employ hard hazing tactics such as shooting bean-bag rounds, rubber bullets, and cracker shells.

The goal of hazing bears from developed areas and educating visitors about proper food storage is to reduce the chances of a bear obtaining a food reward. When bears begin to consume human foods, they often become "food-conditioned" and actively seek out anthropogenic sources. This type of behavior can become dangerous as the bear becomes increasingly bold.

"We don't tolerate food-conditioned bears in the park and across the Greater Yellowstone Ecosystem due to the human safety threat and to avoid bears propagating this learned behavior across the population. Visitors play a critical role in preventing bears from becoming food-conditioned by securing food, coolers, backpacks, and other bear attractants."

In contrast to food-conditioning, bears in national parks are likely to become habituated by losing some of their natural wariness of humans.

"We don't necessarily look at habituation as a good or bad thing, it's what happens when a bear or any other animal is repeatedly around millions of park visitors each year. But when the bear becomes food conditioned, then it's a safety threat," Schwabedissen explained. "The message shouldn't be, 'We're going to go in and take action against the bear,' the message should be, 'What can you do for the bear? What can you do to secure your stuff so the bear will not receive a food reward?'"

"So, what can people expect?" I asked.

"Unless you have a telephoto lens, you likely won't get that close-up of a bear's face, and bear viewing in Grand Teton or Yellowstone will generally consist of standing shoulder to shoulder with a couple hundred people," Schwabedissen replied.

I asked how visitors can help with bear management.

"If visitors encounter a grizzly during their visit, they can help bear biologists by filling out a bear report at one of the visitor centers, especially if they encounter a sow with cubs, come across a bear acting strangely, or stumble upon a bear accessing human stuff."

In addition to helping park staff proactively respond before human-bear conflict occurs, such reporting also contributes important data to the Interagency Grizzly Bear Study Team's integrated population model, which combines field observations with data from numerous other sources such as flight surveys and trail cameras to arrive at annual population estimates.

As we drove past the Colter Bay Visitor Center and down to the swim beach on the rocky shore of Jackson Lake with snowy Mount Moran in the distance, Schwabedissen and I realized we were both at the Human-Bear Conflict Workshop in Lake Tahoe and spoke about the importance of messaging. Schwabedissen said the National Park Service works hard to ensure the messages regarding grizzly bears are consistent, right down to the "Be Bear Aware" signs at the trailhead.

Along with carrying bear spray and making noise, one of the messages Grand Teton is working on is what to do if a bear walks into your campsite or picnic spot. Along with grizzlies, there's a big population of black bears in the park that like to wander around String Lake, Jenny Lake, Phelps Lake, and Signal Mountain.

"We don't want that bear to come in and scare people off a picnic table, because then it just learned that all it has to do is walk in and it gets a free meal. We will ultimately have to remove that bear because it's really hard to break such bold behavior once a bear learns to associate people with food," Schwabedissen explained.

To prevent that, Grand Teton has installed over a thousand large-capacity bear boxes at campsites and recreation areas like Swim Beach and String Lake, allowing visitors the opportunity to do the right thing in a very easy fashion. Ideally, people will be aware they are visiting bear country and have a plan about what to do with their food and belongings if a bear appears.

"And this should be preplanned?" I asked.

"Exactly, people should decide who will be the designated person to stay with the things or have a plan to take their stuff back to their vehicle or put it in a bear box before they jump into the lake," Schwabedissen said.

From the swim beach we followed a roundabout, stopped at the bathroom, and then reversed our journey back toward the visitor center in Moose. We hoped to stop at any bear jams, but as we headed south, all we saw was the ursine paparazzi parked at pullouts, waiting for the next bruin to appear.

"My title may be bear biologist, but ninety-five percent of what I do is manage people," said Schwabedissen, gazing at some tourists venturing into the woods with their cameras. "The more proactive we can be in educating visitors about food storage and bear safety, the less likely we are to have to take action against the bear."

Since much of the Greater Yellowstone Ecosystem has reached its carrying capacity of brown bears—as evidenced by adult males fighting to the death and sows having their first litters later due to more competition and many bears extending their range outside the ecosystem— Schwabedissen believes the biggest challenges will be found outside the protected boundaries of national parks and forests.

"We've made tremendous progress when it comes to grizzly bear recovery, but the forefront of grizzly bear conservation and management is now centered in these local communities and private lands on the periphery of the Greater Yellowstone Ecosystem."

When Justin dropped me off at the visitor center, I thanked him for his time and decided to rise early the following morning and attempt to see 399 and Spirit.

After standing on the roadside at the bridge for half an hour the next morning, a grizzly slowly emerged from the forest.

"Is it her?" I asked a man to my right.

He shook his head. "Too small. Probably a subadult."

Just then, another scrawny bruin steps out from the shadows.

"Definitely two subadults," the man explained. "Could be Bear 926's two two-year-olds. Likely got kicked out by mom and are hanging together until they learn how to survive on their own."

While I'd hoped to spot 399 and Spirit, it was still fascinating to watch the two bears forage along Pacific Creek.

After twenty minutes, the bears disappeared into the woods, and I vowed to keep returning throughout the summer to spot the elusive Matriarch of the Tetons.

Chapter Fourteen

Dancing with the Bear

We're training bears. We're teaching bears what the rules are.
I call it "bear shepherding." We're shepherding these bears
through their lives and trying to help them survive.
—*Carrie Hunt*

Kevin Grange

On a warm July morning in 2022, Nils Pedersen awoke in his tent before dawn and squeezed into his wildland firefighter gear—flame-resistant army-green pants, a yellow brush shirt, and heavy black leather boots.

Nearby, dozens of wildland firefighters from Black Hills, South Dakota, and Whiskeytown, California, were sound asleep. They'd spent the last few weeks working long days and fighting multiple forest fires east and west of the Dalton Highway in the Yukon, two hundred miles north of Fairbanks. The inferno, named the Dalton Creek Fire Complex by officials, had burned thousands of acres, and it was their job to ensure lives, homes, and more forest acres weren't lost.

Flicking on his headlamp, Pedersen hurried over to his truck and opened the tailgate to find his three Karelian bear dogs, Mardy, Rio, and Soledad, waiting with wagging tails.

"Morning, guys," Pedersen said, giving each pup a big scratch.

Hours earlier, at 1:30 a.m., Pedersen and his dogs had prevented a large black bear from entering camp. Soledad had scented the bruin first, waking Pedersen with a loud bark, and then he'd used Mardy and Rio to chase the bear out of camp with a chorus of high-pitched barks.

Wiping sleep from his eyes, Pedersen quickly fed the dogs and attached their chest harnesses and leashes, and then the three-pack headed out into the surrounding forest. Their job? Patrol around camp and ensure no bears, wolves, or moose were in the area, preventing any surprise encounters with firefighters when they woke up.

As thousands of wildland firefighters flock to woods each summer to fight forest fires raging across Alaska and the western United States, it creates the perfect storm for human-bear conflict. There's the opportunity for surprise encounters with black bears and grizzlies; large camps with lots of food and garbage sit in the middle of bear country, and firefighters are often too busy and exhausted to secure attractants. Previously, wildland fire crews would leave a radio on a picnic table with the volume on high in an effort to scare off bruins and other large game—which wasn't very effective—but now federal agencies are increasingly using Karelian bear dogs, a breed of scrappy, smart, and fearless dogs originally from the Karelia region of Finland and bred to hunt large game.

When Pedersen and his dogs didn't find any threats on their morning sweeps, they returned to camp to find the firefighters awake, brewing coffee and cooking breakfast. Pedersen had a busy day planned: discussing bear safety with a crew that had just arrived in the area, distributing bear spray and teaching them how to use it, ensuring garbage was secured and the electric fences surrounding three separate food caches were working, responding to any bear activity, and conducting additional patrols with his dogs at lunchtime, dinner, and when the firefighters went to bed to allow them a good night's sleep without having to worry about bears. Over the course of the thirty-eight days he and his dogs would spend on the fire, he'd document over a hundred and thirty cases of large and potentially dangerous wildlife such as bears, wolves, and moose near the firefighters; he'd deter black and grizzly bears from entering camp on forty-one occasions and actively push bears out on foot with his dogs on ten occasions.

As for the firefighters, they were about to spend the next sixteen hours fighting flames, felling trees, and dodging widow makers—giant dead snags that can fall, crush, and kill. But before all that, the firefighters needed breakfast and a very important morale booster.

"Come here, Soledad!" a firefighter named Amy called out with a wide smile. "Have a treat!"

What puts fear into the heart of a half-ton, apex predator whose scientific name occasionally ends with *horribilis*? Turns out it's a medium-size dog, standing two feet at the shoulder and weighing less than sixty pounds.

I first learned of Karelian bear dogs (KBD) at the Rohrer Bear Camp on Kodiak Island when I met Jeanne Shepherd's pup, Tula. She was a sweet two-year-old, covered in an Oreo-cookie-colored swirl of black and white, who didn't mind squaring off with the behemoth bears.

Shepherd has lived a subsistence life of gardening, hunting, and fishing for the past forty years in some of the densest grizzly country on earth. I was drawn to her story because it is a microcosm of what is happening all over the western United States as farmers, ranchers, and private homeowners are encountering grizzlies who have expanded their range.

"It's tough coexisting with bears because they do what they want. They also live on the Kodiak National Wildlife Refuge, and they're protected, and we're really the ones intruding on them," Shepherd told me as we toured the elaborate gardens surrounding her home. The elevated plots and greenhouse had a voluminous selection of fruits and vegetables—potatoes, onion, turnips, beets, kale, rhubarb, carrots, and raspberries.

Over the years, Jeanne has lost goats and chickens to bear predation. She couldn't smoke salmon on her property or continue to use fish as compost. Grizzlies had knocked down her fences and turned over trees in her property, and the twenty-minute walk to work at the Rohrer Bear Camp on a densely forested path each dawn and dusk was a white-knuckled affair.

"Before I had KBDs, if there was a bear in the yard, the black Labrador I had at the time would run up to the house and scratch the door to go in," she explained. "I was a major food source in September, and a sow with three cubs would move into my garden and stay there."

Along with sows, the other culprits seemed to be subadults.

"When they get run out of the den, they use me as a buffer from big bears because big bears would kill them," Shepherd explained, handing me a carrot that had the sweetness of candy. But fifteen years ago, she got her first KBD, and everything changed.

"If you live in the bush, this is the dog for you," a friend at Alaska Department of Fish & Game told her. "That's their job, they run bears out of the yard."

Shepherd's first KBD was named Yuli, and he could chase an eight-hundred-pound bruin out of the garden with his high-pitched bark and then turn around and calmly play with kids. Yuli had perfect recall and was trained to push bears out of Shepherd's property a hundred yards, without injuring the bear or himself, and then immediately return home.

"My yard is my yard, and the bears know that," Shepherd explained. "And the wildlife refuge is the bears' yard, and my dog gets that."

Along with clear boundaries, Jeanne also adapted parts of her life to reduce human-bear conflict: she picks her fruit and vegetables the moment they ripen, lest they sit on the stalk and tempt bears, and she follows a strict schedule of when she travels to the bear camp.

"I walk the trail at the same times to cook breakfast and dinner. If I go off schedule, I'm more apt to run into a bear because he's got my clock down, and he's not going to be on the trail at that time."

In the years since, the only time Shepherd has had trouble was when she and Tula traveled to the town of Kodiak for two days and left the property unattended. A bear broke into her garden and tore down the fence.

"How do the dogs not get injured?" I inquired.

"Karelians have a way, they just kind of tuck their hind ends when they run, so it scoots out of the way. They're just super agile and quick, and if a bear swats at them, they've already got their tail tucked and they can dodge like crazy."

"Like two boxers in the ring," I replied.

Jeanne laughed and said the bear population around her home has boomed over the last forty years.

"They say in Africa everyone wants to go see elephants, but the villagers dislike elephants because they're so destructive, and that's the way I sometimes feel about bears, really."

Seeking more information about Karelians, I hopped on a plane from Anchorage to Fairbanks to meet Nils Pedersen and visit the Wind River Bear Institute. I first met Pedersen at the Human-Bear Conflict Workshop in Lake Tahoe, where he'd brought his dog Soledad and had an information table set up about KBDs. When he invited me up to Fairbanks to learn more about the breed, I eagerly accepted.

The Wind River Bear Institute (WRBI) was founded in 1996 in Florence, Montana, by an inspiring wildlife biologist named Carrie Hunt. After watching three grizzlies she'd been studying at Yellowstone National Park die due to conflict outside the park's protected boundaries, she decided to do something about it. Hunt pioneered such deterrents as bear spray and nonlethal rubber bullets, but when she heard about a little-known breed of black-and-white, wolflike dogs from Finland, she'd found her calling.

"We can train wildlife just the way we train our dogs," Hunt explained. "Problem bears have lost their wariness of people because humans have taught them that's okay, and we have to teach them it's not okay. We are teaching bears what the rules are."

I met up with Nils one sunny afternoon in May on the scenic campus of the University of Alaska Fairbanks overlooking the Alaska Range and Denali in the far distance. I parked near UAF's Museum of the North, a modern building whose impressive curving white architecture was designed to resemble a glacier. A moment later Pedersen pulled up in his white Dodge Ram 1500, part-truck, part-kennel on four wheels.

"Good to see you again," Pedersen said, hopping out. He was dressed in his wildland firefighter uniform, green pants and yellow shirt, and had spent the morning preparing for the 2023 summer season by taking a work capacity physical test and sitting through refresher classes on weather patterns, fuel loads, and forecasts for the upcoming year. Since I also occasionally fought forest fires with Jackson Hole Fire/ EMS in the Greater Yellowstone Ecosystem, home to nearly a thousand grizzlies, I was particularly interested in this new use for KBDs.

As Pedersen opened the tailgate, a small pack of KBDs spilled out. Their ears shot up like the twin tails of an F-22 fighter jet, and they immediately began following scent trails around the grounds. One dog quickly found a sprawling whale bone, part of a landscape architecture exhibit, and another sniffed around the foundation of the museum.

"He's probably smelling the specimens inside," Pedersen explained. "KBDs can scent a bear a half mile away or even out of my truck, traveling sixty miles an hour. Many times, I'll drive through a neighborhood and the dogs will stick their heads out the window to locate a bear."

Pedersen called the dogs back and we set out on a hike named Pooch Trail. After only a few steps, we were swallowed up by dense forest, home to moose, black bears, and grizzly bears. Despite not carrying bear spray, I wasn't the least bit afraid since we had multiple four-legged bear-detection "devices" patrolling around us. As the dogs dashed around the woods, it was clear they loved to hunt and had no shortage of energy.

Pedersen grew up around working canines in Fairbanks. His parents ran sled dogs, and he recalls a time when his grandfather, Einar Pedersen, showed him a photo of a Husky running off a polar bear, pointed, and said, "He's dancing with the bear." Pedersen received his master's in wildlife biology and conservation, but he knew working with dogs and bruins was in his DNA.

"People don't always appreciate how intelligent bears are," he said as we walked. "People know that bears in captivity are trainable, but they don't apply that same understanding to bears out in the wild."

"So, you believe wild bears can be taught to respect human space?" I asked skeptically.

Pedersen nodded. "A bear in the wild can learn to practice a foraging strategy that exploits the availability and abundance of human-associated foods or, by that same measure, bears can learn to avoid people and maintain a healthy respect for human-occupied space if they are denied access to human foods and conditioned to be wary of human presence. I don't think people know how plastic bear behavior is and how human activities can mold bear behavior in places where we interface."

Like Carrie Hunt, Pedersen's interest in finding a nonlethal way to deal with grizzlies stemmed from the unfortunate death of a bear. He was out hiking with friends who had their dog off leash and out of sight. When the dog encountered a brown bear, he ran back toward the group, bringing the bruin with him. Pedersen's friend fired a warning shot to stop the bear's charge, but the grizzly kept coming, and he was forced to shoot it.

Pedersen first heard about KBDs in 2010 when he adopted a puppy from a little shelter in Fairbanks. "I was working for the Alaska Department of Fish and Game doing fish work on the Yukon and quickly gained an immense appreciation for having a good dog around when working out of fish camps in bear country."

When that dog passed away unexpectedly a year later, Pedersen found himself in Montana searching for another KBD, and he met Carrie Hunt. Nils immediately fell in love with Carrie's dogs and decided to devote the rest of his life to this breed.

"The Karelian breed is a really neat mixture of traits that I value in a dog," Pedersen said. "They are very friendly toward people but very aggressive toward large and potentially dangerous wildlife. They are very healthy, medium-size dogs with distinctive markings and an athletic build. They are intelligent but their intelligence is primitive, which is to say that they have a mind of their own and are capable of independent thoughts and decision-making. This kind of intelligence allows them to hunt and bay up large game effectively without getting themselves injured or killed in the process. This intelligence can make them difficult to train in strict obedience, but they are very loyal to their human partners and, if treated fairly, develop a deep bond with their people. I have a great appreciation for the intricacies of their nature. I find that their strengths are complementary to my own, and I get a great deal of satisfaction working with a dog that I trust and consider to be, in many ways, my equal."

Since then, Pedersen has worked with his KBDs on the three species of bears in North America—black, brown, and polar—as well as the Asiatic black bear in Japan and has helped test, select, train, and place over forty-three of Wind River Bear Institute's dogs with individuals and public safety agencies.

The process of training a KBD begins almost the moment puppies are born. Hunt and Pedersen begin socializing the KBD puppies at three weeks and then test all of them around eight to ten weeks for their working aptitude and finding the right personality match for the committed owners. Based on a puppy's performance, it becomes clear whether the pooch is geared more toward bear conflict, bear protection, or as a companion.

Bear Conflict level dogs are placed into Wind River Bear Institute's "Wildlife K-9" training program for dogs and handlers. Wildlife K-9s are used for detection, human-wildlife conflict prevention & response, and public education. Key uses of Wildlife K-9s include finding bears, even in the dark or in areas of poor sight lines, tracking & trailing, "strike" scent detection work from inside a vehicle, bear den detection, finding carcasses, finding dead or wounded wildlife, responding to mauling incidents, investigating wildlife related crime, functioning as additional personnel to push bears out and away from human use areas, hazing and aversive conditioning, and to be turned off-leash as a non-lethal last resort to harass bears unwilling to leave. Additionally, Wildlife K-9s function as "wildlife ambassadors" to form bridges between people and wildlife through public education and outreach events in which the dogs help deliver important bear safety messages to people in a way that is entertaining and useful.Bear Protection dogs are placed with people that live in bear country and need a dog that will stand their ground in a bear encounter, defend your life and property from bears, but will not be as driven as Conflict dogs to leave you to range out and hunt. Companion dogs are placed with people that like KBDs and want a well-bred, well socialized dog that is not as driven to hunt large game.

The puppies chosen for Bear Conflict work are taken into the field between ten to fourteen months to work alongside seasoned KBDs to learn how to track, trail, detect scents, and be comfortable around firearms and travel in off-road vehicles, boats, helicopters, and airplanes.

Carrie Hunt operates Wind River Karelian Bear Dog Partners in Florence, Montana, and Pedersen runs Wind River Bear Institute in Fairbanks. Typically, WRBI Program Members responds to over eight hundred incidents annually, working with federal, state, and provincial agencies along with private landowners. Despite the many encounters, neither they, their dogs, nor the bears have ever been injured.

"Why can't we just relocate the bears?" I asked as we started hiking back toward the truck.

Pedersen told me relocating bears is costly and time intensive, there aren't a lot of places to put the bruins, and it's not setting them up for success. The bears are taken out of a landscape they're used to finding food in and placed in a different ecosystem with other bears. "Plus, it doesn't address the root problem, which is often people learning to secure attractants like food and garbage," added Pedersen.

"And the bears would probably beat the biologists home," I continued, remembering a story I'd read about a relocated black bear that traveled four states—and over a thousand miles—to return to his home in Great Smoky Mountains National Park.

Relocating nuisance bears usually led to the bear doing the same thing in someone else's backyard leading folks to wonder why a wildlife agency would put this bad bear in their neighborhood. "It can lead to conflicts between people, agencies, and jurisdictions," Pedersen added.

Wind River Bear Institute works their dogs almost exclusively on-leash. "We are usually working in populated areas, and we cannot have dogs chasing bears around off-leash. Its not safe for the dogs, for the bears, or for the people," said Pedersen. "The dogs are being used for very specific tasks as dictated by the handler to respond to and prevent human-bear conflicts."

The dogs are trained not to pursue bears beyond a certain distance; they are handled on-leash and very rarely turned loose on bears. The technique is effective because the KDBs do not pursue bears into "good" bear areas, places that wildlife officials want bear to be.

"We are pushing bears out of the "bad" bear places but once they move into the good bear place, we relieve pressure. This is a negative reinforcer that works to communicate to the bear that this is a good place for the bear to be but every time that it comes into human-use areas we are applying pressure, applying aversive stimulus again until they leave, at which point we relieve pressure again, and repeat," added Pedersen, "The more consistent we are the more effective the technique is. the more that the bear attractants are secure from bear access, the better. It is crucial that food garbage, bird feeders, apiaries, pet/livestock feed etc. is secured from bear access in order for this technique to work."

Some of Pedersen's most rewarding work has been performing polar bear and grizzly bear den detection surveys in the North Slope oilfields of Alaska to help oil companies avoid disturbing denned bears while conducting off-road activities like seismic surveys or ice road construction. Some of the less pleasant but meaningful work Pedersen has done includes locating a dead bear for a hunter on the Alaska Peninsula. The hunter who wounded the grizzly searched for a day and a half without success. Pedersen and his dogs located the bruin in less than an hour.

Pedersen views wildlife K9s like KBDs as "tools" in the management toolbox that are appropriate to use in certain cases. "Humans and bears have a kind of dynamic tension in our relationship, and it is good that we maintain that kind of balance—bears respecting and avoiding human-use areas and people respecting and avoiding bear-use areas," he said.

"Respect the bear and the bear will respect you," I replied, still not knowing its full meaning.

Pedersen nodded. "Maintaining this balance, teaching bears where the boundaries are, and teaching people how to live and recreate in bear country responsibly, that's what the concept of bear shepherding is all about, teaching bears and teaching people."

As we arrived at the truck, Nils loaded up the dogs and said it's common not to see bears while his dogs are doing strike work and sweeps in the forest. "That's the great thing about using the dogs; they can smell the bear and bark, so by doing passes with them in the truck or on foot, we can develop a map of roughly where they are."

Occasionally, Pedersen and his dogs will perform a "hard release," where the KDBs chase a problem bear that has been trapped, and such work is highly nuanced.

"It involves trapping a bear and releasing it on-site. When the bear is released from the trap, it is hit in the butt with bean bags and rubber bullets fired from a 12-gauge shotgun and then, once it reaches half of the distance to a good climbing tree, we release the dogs to tree the bear. Sometimes we just keep the dogs on-leash and pursue the bear, especially with grizzlies that don't often climb trees," Pedersen explained.

The point was to give the bear a negative experience to associate with its behavior and the location and part of that negative experience involves pursuit with dogs.

"Most importantly, we are not turning our dogs loose to chase bears around," Pedersen explained, "We are applying the dogs in a very specific way to accomplish the goal of reducing the frequency and severity of human-bear conflicts. All the human-wildlife conflict work we do come down to one thing—it's not the KBDs that are the solution, it's people. It's education and prevention."

Were grizzlies teachable? Could they be trusted? Was coexistence possible? Would I ever lose my irrational fear of brown bears?

Those were just a few of my remaining questions as I packed for my trip to McNeil River State Game Sanctuary. I'd hoped to bring Meaghan along, but when we weren't selected in the general lottery, I applied for a special access permit for my book and had been accepted.

"What a visitor sees at McNeil is a phenomenon unique in the animal world," John Craighead, the famous bear biologist, had said.

"If you don't visit a place like McNeil, you're missing a huge piece of the puzzle," added Joy Erlenbach at the Human-Bear Conflict Workshop in Lake Tahoe.

At McNeil, I'd confront my biggest fear by sleeping alone in a tent in an open camp without an electric fence among the largest gathering of brown bears on the planet.

As I hopped aboard the floatplane in Homer, Alaska, on June 22, I decided the sanctuary sounded amazing, unique—and totally insane.

Chapter Fifteen

Into the Sanctuary

Bears are not hard to live with; but you have to learn to trust them.
My general rule is that if the bear's not stressed, I'm not stressed.
It's all about the nature of bears and the education of people.

—Larry Aumiller[1]

Jennifer Smith

1. Larry Aumiller, as quoted in *In Wild Trust*. Fair, Jeff, and Larry Aumiller. (Fairbanks: University of Alaska Press, 2017), 73.

I'd always been taught that when facing a bear I should stand tall, make myself appear big, and wave my arms, but now as two young grizzlies approached our group on the sedge meadow, Alaska Department of Fish and Game (ADF&G) biologist, Beth Rosenberg, told us to huddle together, stay silent, and kneel.

Our group of six gathered close, taking a knee as two yearlings with fluffy almond-brown fur paddled toward us like kindergarteners, with shy glances and sudden starts and stops. Suddenly, their mother—a medium-size sow with light brown fur and a dark muzzle Rosenberg had affectionately identified as Bearded Lady—peered over at us.

This could get bad, I thought.

But a moment later, Bearded Lady let out a long sigh, looked away, and lowered her head to rest on her paws.

I couldn't believe it—we were babysitting two young brown bears as their mother napped!

"This female trusts us because we're acting in a predictable manner," whispered Rosenberg.

The yearlings continued forward a few paces and then stopped ten feet from us. Rising to stand, they took a long sniff before dropping back down to all fours, racing back to mom, and play wrestling in the tall grass.

"We were communicating with those bears with very subtle, nuanced body language in a way that they understood," explained Rosenberg, standing. "We were using the language they would use with another animal."

"Amazing," I replied, grabbing my backpack and joining her.

"Isn't it fun to just be another animal on the landscape and not be number one?" Rosenberg said with a wide smile.

McNeil River State Game Sanctuary was created in by the Alaska legislature in 1967 to protect two hundred square miles around Kamishak Bay and provide permanent protection for brown bears, other wildlife, and their habitat and continue the unique bear-viewing experience within the sanctuary.

Located 250 miles southwest of Anchorage, only ten people are allowed to visit McNeil every four days, and the sanctuary promises the closest most openly intimate human observation of *Ursus arctos* on planet earth. The lucky few are chosen through a fair and equitable lottery where the chances of success—depending on the time block one applies for—hover between 2 and 18 percent. When I applied for a Special Access Permit—a "Sci-Ed Permit" for short and which is separate from the lottery—I was invited to visit McNeil during the June 23–26 block.

From Jackson Hole, I flew to Anchorage, then caught a bus to Homer, traveling down the magnificent Kenai Peninsula. During the lecture at the Human-Bear Conflict Workshop, biologist Jay Honeyman spoke of the need to manage at the landscape level beyond jurisdictional boundaries, and nowhere was this idea put into action more successfully than the Kenai-Russian River area, where multiple agencies— Chugach National Forest, US Fish and Wildlife Service, Kenai National Wildlife Refuge, Alaska Department of Fish and Game, and Kenaitze Indian Tribe—all work together to reduce human-bear conflict on a river system teeming with hundreds of anglers and bears.

After a stop at Copper Landing, where everyone on the bus grabbed an ice cream at Wildman's Country Store, we continued south and arrived in Homer, the halibut capital of the world, at 7 p.m. My first order of business was to walk to Ulmer's Drug and Ace Hardware to pick up a can of bear spray. It was cloudy and rainy, and I wondered if I'd be able to fly to McNeil the following morning, but in Alaska, the weather changes so often it does little good to plan anything before ten minutes from departure.

I found my bush pilot, Jimmy Christensen, fueling his plane the following afternoon when I arrived at Kachemak Air Service for my flight. Christensen was an athletic guy in his late thirties, and I trusted him immediately based on his uniform: hiking boots, Carhartt pants, flannel shirt, fleece vest, and ball cap. During my trips to Alaska, I'd realized you want a floatplane pilot who dresses exactly opposite of one who flies for a commercial airliner. If you ever see a bush pilot sporting a suit and tie or polo shirt, run for the hills. Instead, you want the pilot in jeans and a sweatshirt who looks as if he just finished fishing or skinning an elk.

"It's pretty cloudy and windy, so we might have to turn around, but we'll take off and see how it goes," said Christensen, who appeared to be using his truck's motor to power the gas pump, which in turn delivered fuel to his red-and-white Cessna 206 seaplane.

If Willy Fulton on Kodiak Island was the wise veteran, the Chuck Yeager of floatplane pilots who could fly anywhere, anytime, in virtually any weather, then Jimmy Christensen was the young Top Gun. Christensen first flew solo at age seventeen and has been flying commercially since age nineteen. Over the course of his twenty-plus-year career, he's logged more than thirteen thousand hours of flying through the wind, rain, fog, sleet, and snow in the Gulf of Alaska, Cook Inlet, and Shelikof Strait.

In 2020, Christensen took over Kachemak Air Service from Bill and Barbara de Creeft, who'd run the business since 1967. Today, Kachemak Air offers sightseeing tours to the Kenai Fjords and surrounding icefields and to the famous Tutka Backdoor Trail, charter services to hunters seeking to be dropped off in the field, and flights to McNeil.

We loaded my gear onto the floatplane and were soon motoring across Beluga Lake before lifting into the air and leaving whitecaps in our wake. It was a partly sunny day in Homer, and as we flew south along the wilder side of the Cook Inlet, the clouds grew thicker and lower, and Mount Augustine, the most active volcano in the eastern Aleutian Range, hid like a shark fin somewhere in the gloom.

As Christensen flew, I noticed the plane was slowly descending to stay below the cloud deck, and soon we were zooming fifty feet above the ocean as Jimmy casually chatted about a time when both his engines failed and he was forced to land on a crooked glacier.

"As we got close, the client looked at me and asked, 'Are we going to crash?' and I said yes!" Christensen recalled.

"What happened?" I asked.

"We touched down on the snow and I thought, *I'm actually going to stick this landing*, but then the floats hit a crevasse and we flipped, but luckily no one was badly injured."

I appreciate that Christensen was honest with his clients and saved their lives that day, but his stories didn't calm my nerves.

"Congratulations," he said, gazing over at me.

"For what?" I asked.

"For making it out here," he explained. "Apart from the bears, even making it to McNeil Sanctuary is a major achievement. The odds of getting a permit are slim. People can be marooned for weeks due to weather and miss their allotted time slot, and the tides must line up too."

"Thanks," I replied as we followed the rugged coastline due to the low visibility. At this point, I felt as if we were flying by braille. Suddenly I spotted a few small cabins by the beach, surrounded by high bluffs and miles of primeval wilderness—McNeil River State Sanctuary.

"That's it?" I blurted out. The camp had such a small footprint, it looked as if a sneeze could blow it away.

"Yup," Jimmy replied, circling camp. "Big bear on the gravel spit," he said, pointing. "Another one on the edge of the meadow."

As we landed on the lagoon, the staff at McNeil was there to greet me and the other new arrival who'd arrived on a separate seaplane—a retired bird biologist named Dorsey who'd worked at Crater Lake most recently. As I hopped off the plane, Rosenberg greeted me. "Welcome to McNeil," she said. "Let's make a daisy chain."

A daisy chain was a way to move gear and equipment from a float-plane to shore, and it took immediate priority over warm welcomes and introductions. Dressed in our hip waders, we lined up at even intervals and the baggage shuttle began. As Jimmy handed me my backpack, I tossed it to Rosenberg who gave it to Nick who passed it to Tim who lateraled it to Matthew who placed it into a black utility wagon with heavy-duty, off-road tires.

Nick, Tim, and Matthew were all other ADF&G staff who work at McNeil and had years of experience working alongside brown bears at places like the Alaska Zoo and Brooks Falls. Matthew, a bespectacled man with a five o'clock shadow, had also worked at Round Island, a sanctuary consisting of seven small, craggy islands in northern Bristol Bay that protects one of the largest haul-out spots for a large beast with downward-pointing tusks: *Odobenus rosmarus divergens*—the Pacific walrus.

From the lagoon, we dragged our wagons across the sand, admiring Kamishak Bay. Every time Dorsey or I inadvertently veered away from the group—entranced by the beauty surrounding us—Rosenberg or one of the guys would say, "Hey guys, let's pack it in tighter."

Along with monitoring this large concentration of bears and taking small groups out among them, the McNeil staff also ensured visitors followed the strict policies regarding camp conduct. For the next four days, the only time I'd be alone was in my tent or on the toilet. During all other times, we'd be moving as a group, which was safer for us and more predictable for bears.

We arrived at the boundary to camp, marked by a few scattered log seats known as the "mushroom stools," where there were strict rules: we couldn't eat at the mushroom stools, but drinks were allowed. We couldn't stand on the logs for a better vantage point, lest it intimidate a passing bear, and we should never leave camera gear or other belongings on a seat in case a curious bruin mistook it for a toy.

"The way McNeil can work is if we have 100 percent control. We manage humans all the same, and bears expect that predictability," Matthew explained. "They know we're not going to cross this log, and that enables them to come right by. And the bears also know not to come in here."

"We're number two, bears come first," added Rosenberg. "If there's one thing that defines McNeil and makes it unique on the planet, it's that bears come first. It's the defining principle and the way we make decisions. Everything else is easy. If it's good for bears, we do that, and if it's not good for bears, we don't do it."

Our first task at McNeil was to secure all our food and any other scented items—gum, deodorant, sunscreen, toothpaste—in the cook shack, a sixteen-by-twenty-four-foot cabin that would be the center of activity during our stay.

Next, Matthew gave us a tour of the camp. "You're free to roam around, but don't go outside camp boundaries."

From the mushroom stools overlooking Kamishak Bay, the boundary of camp followed a patch of alders until it arrived at a small lily-filled pond and dry-cabin sauna. Inside, water bubbled in a 160-quart heavyweight stockpot perched atop a woodstove. A bucket of cold water, biodegradable soap, and a wooden ladle sat on a wooden bench for bathing.

For thousands of years, steam baths have been an important part of Alaskan culture. "Something bothering you, go take a real strong steam," the Yup'ik elders preached, "then just pour it off." The tradition has continued in Alaska to this day, where a dry-cabin sauna is considered more essential than a TV or a garage. I hadn't planned on showering during my time at McNeil, but I wasn't going to turn down a sauna either.

"You've taught bears these boundaries," I asked, "and they can be trusted to abide by them?"

Matthew nodded. "We *remind* bears all the time where the boundary is and they learn to go around. Every now and then a bear will wander in, but it's rare."

From the sauna, we followed a patch of alders to two outhouses, each filled with bear-related Gary Larson cartoons and a pair of wooden bear claws to hold the toilet paper.

I gazed back at the campsites fifty yards away and asked Matthew the protocol for nighttime trips to the toilet. "Just clap and make some noise as you are moving on the trail," he said nonchalantly.

"Guess I'll be waiting until morning," I replied.

From the outhouses, the boundary of camp led to three small cabins near the beach, where Beth and the team stayed, and from there it was a short walk beside a line of bushes overlooking the sand, back to the mushroom stools. The camp tour ended inside the cook cabin. I tried to make sense of what I'd heard. "What happens if a bear wanders into camp?" I asked.

Matthew reached up to a wooden shelf and handed me a blue Super Blast air horn. "If a bear wanders in, which happens sometimes, just give two quick blasts."

The air horn wasn't to haze the bruins out of camp. Rather, it was to let Matthew and the rest of the team know a bear had wandered in.

"You can bring it to your tent, and it's no problem. If it's 7:00 a.m. in the morning, and people are asleep, still go ahead. Just *toot-toot* and we'll come out and clap the little subadult out of camp," Matthew promised.

"If a bear comes by camp at 7:00 a.m., hopefully I'll be sleeping," declared Dorsey.

Hazing bears at McNeil consisted of using the least amount of stimulus—a soft clap, stomp of the foot, or breaking a twig—to encourage the bear out of camp without spooking it or making it fearful of humans. Outside of camp, we'd never "push a bear," because they came first, and we were guests in their habitat and home.

Along with securing food and scented items, proper handling of garbage was essential, and Matthew informed us there were five ways to handle trash at McNeil. The first white bucket held gray water such as toothpaste or pasta water. "There should be no food particles in the gray water," Matthew advised. "Once the bucket's full, you can go down to the beach, but before you do, look both ways for a bear, and if there are no bears, you can walk to the tide line and dump it."

Each time Matthew said the word *bear*, I grew a little more anxious.

"When the gods wish to punish us, they answer our prayers," Oscar

Wilde once said. I'd wished, hoped, prayed—and submitted a forty-page special access permit application—to experience the largest concentration of brown bears on the planet, and now I was here at McNeil and having second thoughts.

The next bucket held food waste such as orange peels, coffee grounds, or noodles. This small bucket could be dumped by the guides at the end of the spit on a low, outgoing tide when the river channel is swift and deep.

Glass, plastic bottles, and tin cans went into the recycling bin, and two large rubber trash cans lined the back wall—one for burnables like paper towels and the other for non-burnables.

"What about the packaging my freeze-dried dinner comes in?" I asked.

"Non-burnable," Matthew replied.

The trash and the recycling would be carried back to Homer whenever a floatplane had extra room.

Later, I strolled around the gravel tent sites scattered around the cook cabin, trying to jump into the mind of a bruin. I looked for openings in the alder bushes, which framed the camp like a vegetative fence, and pondered where a bruin might sneak through. Finally, I realized I was being ridiculous. There was no way to tell or orchestrate this experience, so I just picked a random site and threw my duffel bag down.

Suddenly, Rosenberg hurried over. "A female we call Quinoa is on the beach, just past the mushroom stools!"

Quinoa was a middle-aged sow with a yearling cub who sat on the edge of the sedge meadow, reclining against a driftwood log, nursing. We hurried over with our cameras and smartphones, snapping dozens of photos.

Later, I crawled into my sleeping bag as if I were some kind of special operations soldier, ready for battle. My headlamp dangled around my neck so it was easily accessible, and I had my can of bear spray near my right hand and the air horn next to my left.

I didn't sleep that night.

Every time the wind rattled my tent fly, I was certain it was *Ursus arctos* trying to enter and tear me apart. Eventually, I just opened my Kindle and read until the sun rose.

Kevin Grange

I must've dozed off at some point because when I checked my watch, it was seven o'clock. Blinking the dryness from my eyes, I performed a paramedic-level detailed physical assessment on myself—*no life-threatening bleeds, chest wall intact, pelvis stable, abdomen soft and supple.* While I didn't sleep soundly my first night at McNeil, I'd survived. "I can do this!" I said, sitting up in my sleeping bag, suddenly excited.

I reached for my water bottle and sauntered over to the cook cabin for coffee. Grabbing one of the blue five-gallon military-style water containers, I filled a tea kettle and set it atop the three-burner propane stove and struck a match.

As I waited for the water to boil, I strolled around the cook cabin that was, at once, a kitchen, library, dining room, and living room. Two picnic tables sat across from a woodstove, above which hung socks and various base layers drying across a white cord. The walls were lined with bookshelves of board games and natural history literature, and many of the books had been signed by prominent authors, photographers, and biologists who'd visited in the past—John Craighead, Douglas Peacock, Seth Kantner, and Doug Chadwick. The late Timothy Treadwell has also visited McNeil and, evidently, hiked around barefoot so he could feel how the sanctuary felt underfoot to the bears.

The Alaska Department of Fish and Game supplied the propane and stove, and all the pots, pans, and cookware were generously donated by Friends of McNeil, a nonprofit that works cooperatively with ADF&G to support the mission and visitor experience; it also protests threats to the bears or sanctuary such as the Pebble Mine project.

Since 2001 mining companies have been pining to build an open pit near the headwater of Bristol Bay to extract gold and copper. The mine would occupy two square miles and reach a depth of fifteen hundred feet and would remove millions of tons of rock annually in the search for precious metals. To assist with the operation and disposal of byproducts, the mining company would also have to build a 270-megawatt powerplant, a 165-mile gas pipeline, and an 82-mile road. According to critics, disposing of the waste would threaten Bristol Bay, home to the largest sockeye salmon fishery in the world, which has fed Alaskan Natives—and bears, and upper Alaska Peninsula habitats—for thousands of years, along with supporting modern-day sport fishing and commercial fishing.

"It'd be a two-hundred-mile gash in the best brown bear habitat in the world," said Drew Hamilton, a wildlife photographer and photo guide who worked at the sanctuary from 2010 to 2015.[2] "People need

---

2. Drew Hamilton, as quoted in Bjorn Dihle, "Pride of Bristol Bay: The bears of McNeil and the Pebble Mine project," *Juneau Empire*, May 23, 2020.

to understand that a McNeil bear is also a Katmai and Bristol Bay bear. They move around a lot."

Dredging off nearby beaches could affect Bristol Bay's abundant salmon run, thousands of which spawn in the McNeil River and Mikfik Creek. Building an industrial facility in the heart of brown bear country would likely increase human-bear conflict, and the road would fragment prime habitat, displacing bears and increasing mortality. Biologists also fear noise from construction and boats moving product might further displace bears. According to the literature, roads equal walls for bears. However, thanks to the efforts of groups like Friends of McNeil, Cook Inletkeeper, and the Bristol Bay Defense Fund—a coalition of tribes, commercial and sport fishermen, conservation groups, and others—on January 30, 2022, the Environmental Protection Agency (EPA) banned the disposal of mining waste into Bristol Bay, citing the Clean Water Act, and effectively ended the Pebble Mine project.

As the kettle whistled, I hurried over to the stove to make a cup of joe and my freeze-dried breakfast skillet of scrambled eggs, hash browns, and crumbled pork patties. Just then, I spotted a bear outside the cook cabin window, nibbling on spring grasses.

I grabbed the air horn and was just about to run outside and blast it twice when I noticed the bruin wasn't entering camp. In fact, the bear, a subadult, was slowly and methodically grazing his way around the camp, honoring the boundaries as if they'd been secured by an invisible electric fence.

Maybe bears can learn to work around us in some ways, I thought. *Maybe McNeil River isn't insane after all.*

Over the course of the next hour, my fellow campers wandered into the cook cabin. Four of them, like me, were at McNeil on a sci-ed permit, and I quickly realized I'd be spending the next few days surrounded by some highly intelligent and fascinating people.

One guy who'd been selected in the lottery was in his late twenties and about to enter Harvard's anthropology PhD program and was researching his thesis. "I'm interested in finding out how people advocate for better lives for bears. What ideas and conception of the world do they have? How does that trickle down to the guest experience here and what the afterlife of ideas are?" he began. "I'm also interested in what can happen to people after spending time here. What people learn from a space like this that's specific. I don't know if there's a scientific receptacle for that knowledge yet and if people who are sitting in a lab or office are going to find it meaningful."

Three others were working on a project to bring artificial intelligence (AI) into the realm of bear research and individual identification.

"The idea is that a picture of a bear and an algorithm will be able to tell you this is Bear 399 or Quinoa," said one of the group. "That way, biologists would have to do less capture and collaring, which can stress bears."

Evidently, a collection of over fifty-five thousand images of known bears—as they've aged over the years—had been taken at McNeil over the last few years, creating a dataset unlike any other in the ursine world.

Two members of the team were computer scientists who worked in a lab in Zurich, Switzerland, and the other was an associate professor of environmental studies who'd received his PhD in zoology and physiology.

At nine o'clock, Beth, Nick, Tim, and Matthew arrived for our morning briefing.

"There are birds circling and active above Mikfik Creek at the rifle," Rosenberg said excitedly. "When we can see that from here, it could mean the sockeye are starting to run." There were two main salmon runs at McNeil. In June sockeye spawned on Mikfik Creek, and in July, the chum swam up the McNeil River.

Rosenberg said to plan on leaving at 10:30 for a two-mile hike to Mikfik Falls through a muddy bog and to be dressed for six to eight hours of bear viewing. "Bring lots of food, water, warm clothes, rain gear, and wear hip waders for the hike," she said, handing us each a lightweight, foldable camp chair with adjustable buckles.

An hour later, we gathered at the mushroom stools, standing in the biting wind and sideways rain with our camp chairs flapping in the gusts.

"There's no such thing as bad weather, just bad rain gear," Nick joked.

It was low tide, and out on the mudflats a lone boar dug for clams with dexterous claws before breaking the shells with his massive teeth and slurping up the salty goodness inside.

Surprisingly, spring is one of the most dangerous times in a bear's life each year. During hibernation, grizzlies can lose up to 40 percent of their body weight and they wake up from a winter famine to a snowy landscape with few food resources. In Alaska, many bears from inland wander to the coast in early spring to search for marine life—clams, barnacles, and the occasional whale carcass—and the early green-up of sedge meadows and salt marshes, offering lush *Carex lyngbyei*. This grasslike herb is a reliable and abundant source of protein, and while it may not be enough nutrition to fully pack on the pounds, it allows bears to bide time until the first salmon appear in the rivers. And both younger bears and females with cubs who cannot compete on the stream or river might rely on sedge for as long as it is available into late July.

From the mushroom stools, we followed a thin trail across the estuary. In the far distance, Quinoa and her spring cub foraged like cattle across undulating waves of green. Gazing out at this untouched, intact ecosystem, everything seemed to pulse with life and energy.

Nick, who'd worked previously in bear management at Brooks Falls in Katmai, led the way with a 12-gauge shotgun slung over his shoulder, loudly announcing, "Hey Bear!" every time the trail visibility was limited.

As we walked, Rosenberg told me about the history of McNeil.

The spot where the cook cabin sat had been a seasonal fish camp for Alaska Natives for thousands of years and some of the semi-subterranean dwellings were still at the edges of camp, grown in with alders and beach rye. Later, Charles McNeil (1859–1948), a trapper and prospector, acquired a claim in the hills to the southwest and built a log cabin at the mouth of the river. McNeil staked claims, hunted seals, and sold bear skulls to pay off his mining debts. Not having much luck finding gold, he eventually moved to California with his wife to raise chickens and farm.

Geologist K. F. Mather surveyed the area in 1923, bestowing the name "McNeil River" on it, and word of this special place reached the wider world in 1954 when *National Geographic* published an article with "15 illustrations in natural colors" titled "When Giant Alaskan Bears Go Fishing." Since hunting was allowed near the area at the time, the author, Cecil E. Rhodes, didn't mention the location, but the story captivated thousands.

In 1967 the Alaska State Legislature voted unanimously to designate the land as a wildlife sanctuary. While the sanctuary protected bears, there was nothing in the founding documents about the number of visitors allowed to visit at one time or appropriate behavior. As a result, the "river of bears" quickly became overrun with photographers and tourists toting guns for protection. When one guest waded across the river, dodging bears as he stepped from boulder to boulder, and set up a blind on a rock in the center of the cascade, biologist Jim Faro realized bears were being pushed off the river and bad judgment was going to result in someone getting hurt. Faro realized he needed an onsite manager at the sanctuary, and in 1976 he hired a pioneering man named Larry Aumiller.

Aumiller grew up in Denver and had no formal training in biology. Instead, he studied fine arts in college and worked as an army illustrator at the Pentagon during the Vietnam War. Once his enlistment ended, Aumiller drove to Alaska, where he had a few different jobs and eventually was employed by ADF&G. He did fisheries work until he was hired for McNeil. When Faro interviewed Aumiller for the job, he knew he'd found the right person. Despite not having a background in science, Larry had the inquisitive mind of a biologist; he was smart and creative, had solid handyman skills, and, most important, had a calm but quirky temperament, which was perfect for helping anxious guests feel at ease around *Ursus arctos*. Together, Faro and Aumiller wrote the management plan, consisting of a lottery that limited visitation to ten people every four days, and guidelines for behavior at camp, on the trails, and at the viewing areas.

"Historically, bears have been considered vermin and were all too frequently misunderstood or demonized," Rosenberg explained. "No one had ever lived with brown bears this way before. Faro was proposing we operate McNeil as if bears could habituate to a human presence if that presence was kept in check and remained predictable. Larry was the guy who figured out how to do it. That in and of itself makes McNeal a unique thing from the beginning. Larry figured out through trial and error what bears are like through their behavior and how they can be communicated with through body language. He observed over time that bears have certain habits and that while individual bears are different from each other, general rules apply. If our behavior remains predictable, bears can learn to ignore us as a neutral presence."

Along with the small group size, Aumiller implemented a system where the humans consistently hiked the same trail at the same time and in mostly the same few places at McNeil Falls or around the sedges and Mikfik Creek. Aumiller and sanctuary visitors never approached a bear, but the bruins were welcome to venture close to wildlife viewers if they chose to as they grazed sedge, played with other bears, mated, or fished.

At the time, brown bears were thought to be only bloodthirsty beasts, and few believed grizzlies and humans could coexist, but Aumiller proved them wrong. Like Jane Goodall's work with chimpanzees in Tanzania's Gombe Stream National Park, Aumiller spent hours in the field observing bears and scribbling down his findings.

"Safer interactions have to do with knowing about bears and knowing individual bears, what their tolerances are, and what upsets them," Aumiller said. "What upsets a lot of bears is not knowing what you are. When bears are comfortable, they stick around, and we can watch them. And when they are comfortable and unstressed, they are safer to be around. So, it turns out, safety also leads to proximity."

Aumiller devised a new standard for studying bears in the wild, long before the webcams at Brooks River popularized the idea of bears as sentient beings and unique individuals with different personalities.

"What's great about the whole system is that it turns out that the safest way to do it is also the way that makes the bears feel most comfortable, which is the way that encourages them to be there in numbers and then creates a scenario whereby you get the best viewing," added Aumiller. "Safety, maintaining numbers and bear-viewing experience are all connected in the same direction. If you do one of them one way, everything else gets pulled along."

Aumiller became an expert on bear behavior. "We trust them, and they trust us. We didn't know that could be, but everything is based on this trust," he said. "They are not unpredictable."[3]

In one summer, McNeil staff counted over one hundred and forty-four different bears along the river, including a record-setting seventy-four brown bears feeding at the falls at once. By the time Aumiller retired in 2005, he'd managed the sanctuary for thirty years and by his estimate had logged over twelve thousand, six hundred hours watching bears, led over six thousand visitors safely through the sanctuary, and had over sixty thousand encounters with wild brown bears.

Despite the abundance of bears in close proximity to humans, no one has ever been injured or killed by a bear since McNeil Sanctuary's creation in 1967, and no bear has ever been relocated or euthanized. It's a coexistence miracle that continues to this day.

---

3. Larry Aumiller, as quoted in *In Wild Trust. Fair, Jeff, and Larry Aumiller.* (Fairbanks: University of Alaska Press, 2017).

In 2009, Aumiller received an honorary PhD from the University of Alaska Fairbanks for his contributions to the understanding of brown bears, and even at age eighty, he still finds time to visit the sanctuary to volunteer or help with research projects.

Rosenberg came across the 1954 *National Geographic* article about McNeil when she was researching an undergraduate paper on the mythology of bears and began corresponding with Aumiller. Years passed, and after several unsuccessful attempts to obtain a permit through the lottery, she traveled to McNeil as a volunteer in 2010, calling it the most memorable outdoor experience of her life. Rosenberg has been back every summer since, working her way up through the ranks to now manage the sanctuary.

"What's the key to managing bears?" I asked as we started up a steep hill.

"We don't manage bears," Beth replied. "The key is managing the people."

The headwaters of Mikfik Creek are high in the glaciers and lakes of the Aleutian Range, and as we arrived at the stream, we were delighted to discover sockeye salmon pulsing upstream through a shallow, fast section of the river known as the riffles. A dozen bald eagles waited in the grass nearby.

"I'm surprised there are no bears here," said Nick, gazing around.

"Must be up at the falls," said Rosenberg, pointing to a flock of birds in the distance. "Look at the bird storm."

None of us could wait to get up to Mikfik Falls, but first we had to hike through "the jungle," a shadowy, dense patch of woods and alders that was, evidently, also a well-worn bear trail at the time of the summer when fish were running in the creek.

"Hey bear!" said Nick, walking forward.

As we entered the jungle, the air cooled and birds stopped singing. The trail became a muddy slurry of dirt and grizzly dung, and it felt as if our group instinctively huddled close together, as if in a haunted house. I kept my head on a swivel, certain a bear would hop out at any moment. Nick stopped the group at a small sapling with several broken branches and a smattering of bear prints pressed in the mud.

"This is a rub tree," Nick explained. Rather than scratching, bears use rub trees as a kind of bulletin board to communicate and to help find mates. The scent laid down in scratching is a marker of who has been here."

"Looks like bears just passed through," I remarked.

"They probably did," added Dorsey.

Following the jungle, we started up a knoll that led to our viewing spot. After fifteen minutes of effort, Mikfik Creek appeared below—a thin, shallow stream winding through the forest. Fish leaped and danced in the cascade as a half dozen bruins struggled to catch them using every available technique—snorkel, stand and wait, dash and grab, and the deep dive.

Rosenberg pointed to a tiny patch of grass at the top of a meadow, at the intersection of two bear trails overlooking the creek. "This will be one of our viewing spots for the next few days," she said, setting down her backpack and opening a folding chair. We all followed suit and, as with hiking, sat all bunched together. The arrangement had multiple benefits: it kept us warm in the cold wind and made us appear as one small unit to the bears, and if a bear was chased off the river by a more dominant individual and running up the hill in our direction, he wouldn't run through the group.

"Good to know," I said, grabbing my binoculars and camera.

"We are catching this spawn on the perfect day," said Nick. "This pulse might be over by tomorrow."

There were so many salmon, the bears were high-grading, or eating only the fattiest parts of the fish. And in the midst of seeking food on the creek, the personality of each bear was on full display. One bear tossed a fish in the air with his mouth repeatedly, as if playing catch with himself. Other bears looked like they were swimming the backstroke in the chest-deep water below the falls, and one old boar had fallen asleep standing up with a half-eaten fish dangling from his mouth. Younger bears moved along the ridge edge, too small or too young to compete with the older, larger bruins.

As I sat with my camp chair in recline mode, I chatted with Rosenberg about the discipline necessary for living with bears. "It's not that we make sacrifices, but we have to adjust our needs and wants in the moment so we can coexist," she explained. "But ultimately if you're willing to have discipline and do what needs to be done, you can live in an open camp right next to the world's largest congregation of brown bears."

I asked what the secret was.

"Nothing that happens here is a coincidence or good luck. It happens because bears are really smart and predictable, and we keep 100 percent control over the behavior of a small human group that includes ourselves," added Rosenberg. "There's nothing Pollyanna here. Bears aren't our spirit animals, and they're not our friends. We're two species who are living in the same place and not competing for mates or food. Over time, both species have learned predictability from each other. Ultimately, bears can depend on our predictability to the extent Bearded Lady allows her two cubs to approach us and she is not concerned about what we're going to do. She knows we won't change our behavior, and we can trust the same with her, based on what we've seen so many times. Cubs are another story." Rosenberg smiled.

I mentioned the wild trust that can develop between humans and animals that Aumiller and author Jeff Fair wrote about in their book *In Wild Trust: Larry Aumiller's Thirty Years among the McNeil River Brown Bears.*

"Exactly," replied Beth. "Bears at McNeil aren't different from other wild bears. They have a limited amount of time to get all the food they need. It's a resource-rich environment out here. And it's way more important they get food versus our seeing them get food. So we think about how to move around in this environment so as not to interfere with the critical activities they're working on in a short amount of time."

McNeil had its own vocabulary, and I learned some of it seated at the Mikfik Falls that afternoon. A *drive-by* was another name for a proximity moment, when a bear decided to stroll right past the group. A *goober* was the affectionate name for a subadult bear. A *goob gang* was a pack of subadults, and a *bruiser* was the term given to a big male.

A scrawny subadult with skinny legs and tall, batlike ears called Dingo was my favorite goober, and according to Rosenberg, the most dominant bruiser in the sanctuary was a big ol' male they called Chops.

"Chops might come around," said Beth, gazing around. "I wonder if there are already some chums in McNeil River."

"I'd love to see him," I replied, not knowing that I would soon, and at very close range.

Later as I crawled into my sleeping bag, I could feel a change beginning to take place inside. Were bears really predictable? Could they learn to live around us and could we trust them? The biologists I'd spoken with had been saying this for months, and I was finally beginning to believe them.

Respect the bear and the bear will respect you, the Native Elders on Kodiak had preached for millennia—the phrase felt like it was finally coming into clarity for me.

I was still afraid of grizzlies. Once again, I had my bear spray and air horn positioned at each hand as I struggled to sleep, but I could feel my paranoia beginning to loosen its hold.

Chapter Sixteen

From Conflict to Coexistence

Bears are made of the same dust as we, and breathe
the same winds and drink the same waters. A bear's
days are warmed by the same sun, his dwellings
are overdomed by the same blue sky.

—*John Muir*

Jennifer Smith

I experienced my first drive-by on my second day at McNeil River State Game Sanctuary. The proximity moment—or *Holy shit! There's a wild brown bear heading toward me!*—occurred during the evening of June 23. Matthew, Rosenberg, the computer scientists, and I were all seated around the mushroom stools, chatting casually, when we noticed Quinoa and her spring cub feeding on the tall grasses lining the logs that formed the boundary between the beach and camp.

"I bet if we stay low, they'll come close," Rosenberg whispered, slowly easing herself off the wooden stump.

Despite Bearded Lady's yearlings ambling toward us the first day, I was still a little unnerved moving down to the ground when a bear approached.

Here goes nothing, I thought, taking a seat on soft dirt.

My second day bear viewing at McNeil had been as amazing as the first, highlighted by the first appearance of the most dominant bear in the sanctuary, Chops. Early in the day the sockeye were still pushing up the river, and we'd spent most of the cloudy morning at upper falls on Mikfik Creek.

"We're just humans observing a world," Beth marveled as a large sow snagged a fish midair. "McNeil affects people differently but profoundly, each in their own way. We don't have to tell people to be affected by McNeil; we just live here quietly as one inconsequential piece of the big system that is just about the right balance."

The sockeye pulse had lessened from the previous day, but it was still interesting watching the dance of dominance between some old boars and mature sows for the best fishing spots.

"McNeil was really the first place to experiment with this idea of coexistence in the format we now think about as bear viewing," Rosenberg explained. "For a lot of other places, with other species, this is a model of wildlife management because of the things Larry learned and pioneered."

When I asked how McNeil has changed over the years, Rosenberg said it hadn't and that this is maybe one of the most remarkable things about it.

"How many places can you think of where humans have been for nearly sixty years that haven't changed at all? This is a place where, once it was established, the footprint of the camp has stayed exactly the same with the few very simple parameters for humans put in place at the very beginning."

Beth repeated a quote by Page Spencer, chief of natural resources at Lake Clark National Park and Preserve. "Page said, 'McNeil is what happens when, for fifty years, someone has the courage to say no.'"

"I love that," I replied.

Some of the many things McNeil has said no to over the years, through its staff and nonprofit, is increasing visitation, expanding the cook cabin, building a lodge, hiring a third-party concessionaire to provide food and beverage, allowing trophy hunting in nearby areas, and acceding to the Pebble Mine project.

"If you wanted me to encapsulate what's unique about McNeil, it's that each of us has taken on the role of stewardship; it's that we just hold the line," Beth said. "The answer is no if it's not good for bears. If it's good for bears, we do that. We've found a way to quietly hold the line, and I can't think of anything more unusual than that."

Just then, Tim—another ADF&G staff member who'd worked with grizzlies at the Alaska Zoo—interrupted us, pointing to the hillside across the creek. "There are two mating pairs to the left."

I turned and spotted multiple courting couples in an overgrown field between the alders. "Is it late for mating season?"

Suddenly, Beth stopped us midsentence and pointed to the river. "He's here!"

When I turned and spotted the bruiser ambling down to the river with a wide cowboy swagger, I didn't have to ask who it was.

Weighing over twelve hundred pounds coming out of hibernation, Chops was a brick of a bear with a huge head and legs like giant Sequoia tree trunks. His body was a thick mass of ripped muscles and rolls of fat, and a shiny rubbed spot decorated his rump.

As he ambled down to the river, the other bears didn't just make room for him or offer up the prime fishing spot—they left the water altogether.

"He cleared the entire river," I said, snapping photos. "He's a bruiser all right!"

Surprisingly, bear viewing that afternoon became less interesting with Chops on the river because no other bears dared step into the water. They'd nap for a bit, then wander down toward the creek on well-worn century trails, and upon spotting Chops, they'd freeze, ponder their options for a moment, and hurry back up the hill.

"That subadult literally shook his head no when he saw Chops!" I said with a laugh.

Around 4:30, we packed up and hiked back to camp, checking on a few trail cameras along the way.

An hour later we were seated on the mushroom stools, and that's when Quinoa and her cub decided to do a drive-by.

As Beth, Matthew, Dorsey, a visitor named Mu, and I assumed our places on the ground, Quinoa glanced at us for a quick second, then continued moving slowly in our direction, followed by her spring cub. As they approached, I heard horse-like mastication sounds as they chewed sedge. I was seated behind Mu, who'd never camped or hiked in her life, I decided she was the ultimate badass. She was out here in the last frontier, sleeping in an open camp with grizzlies, and now she was seated front and center, about to have a proximity moment with a six-hundred-pound brown bear.

"Stay still," Rosenberg whispered as the bruins inched closer.

Surprisingly, I was more excited than scared, a feeling I attributed to knowing signals and signs for a bear in distress, and Quinoa displayed none of them. After months of speaking with biologists and traveling across North America, I'd finally become more fluent in bear behavior.

By now only the log forming the boundary of camp separated us from Quinoa and her cub. And then the most amazing thing happened: when Quinoa was directly in front of us, she stopped, sniffed, then turned her back to us and lay down only a few feet away.

By then my head was about to explode with excitement. We didn't exist to the bears. We were just part of the landscape, and there was nothing like the primal experience of being at eye level with a wild grizzly inches away.

Quinoa's cub followed her, and when she reached us, the sow rose back to her feet and then the pair continued, grazing along the boundary of camp.

It felt good that I had to change when I visited McNeil—the rigid camp boundaries, not being able to wander up through the hills or down to the beach by myself, hiking as a tight group, and the strict garbage disposal protocol. I didn't mind the discipline that came from living in service of something outside myself—the bears and their habitat. I savored feeling small in the big face of nature.

Jennifer Smith

"Here, you have a once-in-a-lifetime experience every day," the acclaimed bear biologist John Hechtel once said of McNeil,[1] and he was right.

On my third day at McNeil, we had another drive-by. This time the proximity moment occurred when a curious subadult sauntered right past us in the tall sedge near the Mikfik riffles. I was gazing in the opposite direction when I turned and spotted the bear silently pacing past the group.

The brief sockeye run up Mikfik was nearly over by the third day. Only a goob gang of five subadults was fishing at upper Mikfik Falls, using the high-energy-low-return tactic of displacing water with a great leap and trying to pin a stunned fish against the mud.

Dingo was the worst at tactics. He would stand tall in the waist-deep water with his large ears, then belly flop with a big splash, always coming up empty-handed. The spectacle was both hilarious and heartbreaking. He clearly had much to learn about fishing if he was going to survive into adulthood. Apparently, this is not unusual for the goobers.

We watched the younger bears fish for hours without much success, and on the way home we had a surprise encounter in the jungle with two big bears, likely a mating couple.

"Hey Bear!" Nick said as he rounded a corner and came face-to-face with a large sow.

Instead of charging, the bears simply cut into the dense alders, went around us, and then hopped back out a few yards down from us. It was yet another example that the tips on how to stay safe in bear country—hiking in a group, making noise, standing your ground, not running when approached, and securing all attractants—really work.

"Like aberrant humans, aberrant animals are rare," Rosenberg explained as we continued hiking. "You'd be amazed how many of the problems most people have in a bear encounter are human errors.

---

1. John Hechtel, as quoted in *In Wild Trust*. Fair, Jeff, and Larry Aumiller. (Fairbanks: University of Alaska Press, 2017), 93.

There's an understanding you have after living in this environment that will translate to other places whether you realize it or not."

On my last day in the sanctuary I hiked to McNeil Falls with Rosenberg and a few others. Rosenberg was putting up a couple of trail cameras at the falls to help with learning more about individual bears. Since the chum salmon weren't running, it was doubtful there'd be any bears around, but I still wanted to see the famed falls.

As we hiked through the sedge meadows, we spotted Chops doggedly following a sow and hoping to mate. We also encountered Quinoa and her cub on the far side of the estuary. Only twenty-five yards between the river and the high bluffs separated us as we approached. I realized we were about to test her trust. The sow and cub had trusted us to be predictable and stay in one place the day before when the family approached us by the mushroom stools. And now, as we moved forward, we'd discover if this concept still applied.

"Go slow and stay in a group," Nick whispered. "If they look up, we'll stop because we don't want to displace them."

We packed in close and slowly proceeded. Quinoa and her cub didn't even stop chewing or gaze up at us. What Larry Aumiller and Jeff Faro wrote about in their book was correct: a sacred, wild trust between humans and animals is possible. Or, put differently, predictability can be trusted.

McNeil Falls was everything I hoped it would be, a beautiful, tumbling cascade singing through rolling hills of bushy tundra with a backdrop of snow-covered peaks.

As we hiked home that afternoon, I asked Beth how McNeil can be a model for us in the Lower 48.

"McNeil proved that by controlling human behavior there's the possibility of predictable responses in a species we have labeled as unpredictable. It can serve as a model on how to contain ourselves a little bit and why that's important. It's about people managing themselves. Bears here are doing what they'd otherwise be doing if we weren't around. We manage the people. That is what can translate anywhere."

As we emerged from the jungle on our hike home, we spotted Chops a hundred yards away, still with his head down, intently following the sow. We continued hiking through the sedge with a tall bluff immediately to our right, when suddenly the female turned and started moving directly for us. Chops immediately followed.

"They're moving this way," Nick said. "Let's pack it in tight."

The sow spotted a tiny trail leading up to the steep bluff behind us and increased her speed.

"Closer," Nick said as the bears approached. "And let's stay still. Don't move."

We were about to have a drive-by with the sow and Chops at a very high rate of speed. Every instinct in me wanted to run, but I just kept telling myself, *Stand your ground. Stand your ground.*

The female, a medium-size sow with a distinctive brown coat, thundered past and climbed up the bluff directly behind us. Then came Chops, barreling immediately after her and struggling up the near-vertical bluff behind us.

And here I was, standing on the trail and looking up. When Chops reached the top of the bluff, all his sexual frustration poured out in an angry torrent. Huffing, he rose on two legs, rubbing his back violently on a tree and snapping heavy branches like toothpicks.

"Let's keep walking," said Nick.

Our group continued, but the sow moved in the same direction on the bluff above. Chops dropped back to all fours and continued his pursuit. Suddenly, the female spotted an opening in the alders and tore down the bluff and crossed right in front of the group.

"Here she comes again," I managed.

Just in front of us and before I could catch my breath, Chops came thundering after her, twelve hundred pounds of teeth, muscle, fur, jaw, and claws, charging down the hill after her.

My heart raced. My arms and legs went numb. Breath left my body. The stress response of fight or flight sizzled through my body as everything morphed into slow motion.

And then, amid my panic, I noticed Chops had his eyes firmly fixed on the sow, and it dawned on me: *He's running toward us but not at us. Stand your ground.*

Earth shook with a seismic earthquake energy as Chops thundered past the group and out onto the sedge meadow, where he slowed to an ambling gait and continued following the female with his head down.

Once the pair was a hundred yards away, we all caught our collective breath.

"Holy shit," exclaimed one person.

"That was intense," I added, winded.

Nick said that this is what we would call a close encounter. "But the important beahvior to note was that Chops's eyes were firmly fixed on her the whole time. She was his intention."

We walked in silence for the remainder of the hike, each of us savoring another magical McNeil moment.

I was still wired when I got back to camp, so I walked to the sauna, changed into my bathing suit, and sat there sweating for half an hour before using the soap, ladle, and bucket of ice-cold pond water to clean up.

After my fire bath, I toweled off and changed in the cool summer air, overlooking the sanctuary. Gazing out, I realized something had changed in me during my time at McNeil. Bears could learn about us and, in turn, be trusted. I also had faith that with just a little discipline and effort, humans could coexist with grizzlies. And, most important, I was no longer afraid of brown bears. I'd always be extremely *cautious* and carry bear spray in grizzly country, but the paralyzing fear had disappeared. As I walked back to my tent that evening, I suddenly knew exactly what the Kodiak Elders meant about respecting grizzlies.

If you respect bears, you'll make the effort to learn about them, I thought. *If you learn about them, you'll begin to understand them.*

When you understand bears, you'll act consistently and safely around them. If you act consistently around bears, they'll act consistently around you and won't react defensively or aggressively, so it comes full circle: respect the bear and the bear will respect you.

Biologist Larry Van Daele had explained this to me as we dined on fish and chips that night at Henry's Great Alaskan Restaurant, but I didn't truly learn it because I hadn't yet lived it.

"Thank you, McNeil," I said, filled with gratitude and gazing out over Kamishak Bay and the surrounding mountains.

Perhaps the greatest part of learning about grizzlies was my travels to amazing places and meeting a host of people—biologists, guides, naturalists, caretakers, teachers, trainers, pilots, scientists and sportsmen—who truly care about our world and are working hard to conserve it. I hadn't expected that my quest to learn about the secret life of brown bears would also reveal some of the best parts of humanity. I resolved to join this passionate tribe working on behalf of grizzlies and get involved when I returned home to Jackson Hole.

Then I crawled into my tent, which sat in an open camp among the largest concentration of brown bears on the planet, and fell into a deep, dreamless sleep before waking up and catching a floatplane flight home.

Epilogue

A Future with Bears

In the end we will conserve only what we love,

We will love only what we understand,

We will understand only what we have been taught.

—*Baba Dioum*

Peter Mangold

On a recent Wednesday in August, I logged on to my computer, navigated to explore.org, and signed in as @kevin_bear_fan. I'd volunteered for the summer to work a few shifts each week as a chat moderator for the live bear cams at Brooks Falls in Katmai National Park. Part of my job was monitoring comments and chats, ensuring everyone was following the community guidelines—no posting any negative comments, no trying to sell anything—but I was primarily there to encourage discussion about grizzlies and help answer questions followers might have about famed bears like Otis, Holly, and Grazer. Along with volunteering as a chat moderator, I helped Jackson Hole Bear Solutions repair bear-resistant trash cans, created a "Living with Grizzlies" and wilderness medicine workshop to teach around town, and hoped to return to McNeil River State Game Sanctuary as a volunteer in the future to help set up camp for the summer bear-viewing season.

"Why is that bear standing on two legs?" asked one follower named @mamabear93.

"Bears often stand to get a better view, or sniff," I replied.

"Why do salmon only seem to jump the falls in one spot closest to the camera? And why aren't bears on the lip of the falls if it's open?" inquired @snowwhitebear.

"From my experience watching salmon and bears at Brooks Falls and McNeil River Sanctuary, the salmon jump up at various places over the falls; however, they often try for the path of least resistance," I typed. "As for the bears, they are spread out along the river based on where they fall in the dominance hierarchy. There might be open spots on the lip or river, but less dominant bears won't venture there."

"It's July 25, where is Otis?" asked another viewer named @UrsusAndrew.

"The arrival of King Otis is probably delayed due to the late salmon run," I wrote.

When I returned to Jackson Hole, I was excited to help people learn about grizzlies and get involved in improving human-bear coexistence. But my highs from visiting McNeil River State Game Sanctuary were quickly tempered as my plane landed and I stepped right back into the leading, bleeding edge of human-bear relations.

Since I started on my journey a few years ago, one bear from 399's litter of four, Bear 1057, was euthanized after leaving the boundary of Grand Teton National Park, becoming food conditioned, and exhibiting dangerous behavior. In April 2023, a five-hundred-pound grizzly was found shot fourteen times just outside the east entrance of Yellowstone National Park, near Cody, Wyoming, likely a victim of vandal killing. Also that month Alaska Department of Fish and Game revived a predator-control program near Dillingham and killed ninety-four brown bears, five black bears, and five wolves in hopes of boosting caribou numbers. And in July 2023, a month after I visited McNeil Sanctuary, the state of Alaska asked the US Supreme Court to reverse the EPA's rejection of Pebble Mine with hopes of kick-starting the project once again.

Fortunately, there'd been some good news too. In Washington State, federal agencies were formalizing a plan to restore grizzlies to the North Cascades, and the public is widely supportive. Brown bears in Montana, Idaho, and Wyoming continue to expand their range and are showing up in areas where they haven't been seen since the 1800s, like the Pryor Mountains near Billings, Montana. There's also strong evidence that brown bears will soon make the trek between the Greater Yellowstone Ecosystem and the Northern Continental Divide Ecosystem, connecting those two habitats. And just one day after followers of the Katmai bear cams urgently asked me when a famed bruin would return to his "office," Otis appeared. "The King Is Back!" newspapers declared to millions of delighted readers. I knew Otis, with his big Buddha belly, would eventually teach us the final lesson of impermanence—but not this summer.

Many great changes have also taken place around Jackson Hole. Following 399's walkabout in Teton County in November 2021, over seventy-five thousand people signed a petition to improve coexistence efforts in Teton County, and local officials listened. Land development regulations were updated to mandate bear-resistant trash cans, dumpsters, and enclosures for all residents and businesses outside of the town of Jackson Hole. In town, residents within the designated bear conflict zone are required to follow the same regulations. To support the efforts, local nonprofits like Bear Wise, operated through Jackson Hole Wildlife Foundation and JH Bear Solutions, an offshoot of Wyoming Wildlife Advocates, are offering bear-resistant trash cans at a reduced price, along with hosting outreach and education programs.

Teton County also updated its wildlife feeding ordinances, requiring all bird feeders to be inaccessible to bears year-round, and mandated no new plantings of ornamental fruit-bearing trees and that existing trees must be harvested or fenced in.

Signage around popular trailheads also improved. "Jackson Hole Is Grizzly Bear Country: Be Prepared," a sign announces at the start of my favorite trails in the Bridger-Teton National Forest. A kiosk in baggage claim at the Jackson Hole Airport now offers bear spray rentals and pamphlets on staying safe in grizzly country. If awareness of a problem is the hardest part of solving it, Teton County is well on its way to becoming a Bear Smart Community.

At home, Meaghan and I have also made the discipline of securing our attractants and garbage from bears a permanent part of our life. To Meaghan's great surprise, hummingbirds floating on invisible wings still flocked to our garden to consume nectar from brightly colored flowers, and a host of other winged visitors splashed daily in our birdbath. Thankfully, it's not just the birds who have returned. As I logged on to my computer this fall, I noticed an email from our HOA. "Hello Sage Meadow Owners," the message began. "A neighbor found bear scat near the bike path on two different occasions. Since we are in active bear season, please be vigilant." Unlike the first time the HOA emailed us in November of 2021, Meaghan and I now relished the thought of *Ursus arctos* and *Ursus americanus* moving through our backyard.

I view my travels to learn about grizzlies as a journey to deprogram my mind from the cult of the killer bear. The quest has given me a wider definition of what being a bear person means. Along with someone who has pledged to keep bears wild by securing attractants like garbage, composts, bird feeders, and beehives around their house and doesn't mind incurring some risk while living, recreating, and working in grizzly country, being a bear person also means I believe in intact ecosystems, predator-prey dynamics, connected habitats, clean water, and abundant salmon runs and supporting all the flora and fauna that fall under the umbrella of this one fascinating apex predator. And when I think of the magnificent creature who set me on this journey, I think of one bear who helped me reconnect with the wildness in me; I think of one devoted mother who filled me with a joyous awe at the beauty and mystery of our planet; I think of the Matriarch of the Tetons. I think of 399.

Peter Mangolds

Acknowledgments

This book would not have been possible without the help of many kind people who encouraged me at every stage of the research, writing, and publishing process.

I am honored *Grizzly Confidential* found a home at Harper Horizon and feel grateful to be included among its impressive list of inspiring titles and authors. For his enthusiasm, encouragement, and belief in this project, I thank my wonderful editor, Austin Ross. My appreciation also goes to Matt Baugher, Josh DeLacy, Kevin Smith, Lauren Kingsley, Hannah Harless, and Andrew Buss.

A huge thank-you goes to my literary agent, Jane Dystel, who doesn't just sell books but builds careers. This is the third book I have worked on with Jane and she—and her business partner Miriam Goderich—always offer invaluable creative ideas, industry expertise, and friendly support at every stage of the sales, writing, and publishing process.

I am particularly indebted to everyone in the brown bear community who generously agreed to speak with me, answered my many questions, and, in many cases, invited me to join them in the field: Charlie Robbins, Chris Servheen, Mike Fitz, Leslie Skora, Shannon Finnegan, Willy Fulton, Jennifer Culbertson, Beth Rosenberg, Adam Dubour, Nate Svoboda, Doug and Lynne Seus, Kerry Gunther, Linda Masterson, Justin Schwabedissen, Tyler Braslington, Jessica Hadley,

Todd Orr, Chris Forrest, John Waller, Joy Erlenbach, Nils Pedersen, Carrie Hunt, the Rohrer Family and their great team of guides (Hiram, Chris, Brett, Jason, and Alan), Don Nelson, Jeanne Shepard, Larry Van Daele, Lauren Sullivan, Dena Hicks, Tammi Hanawalt, Amy Petersen, Heiko Jansen, Joanna Kelley, Ellery Vincent, Jessie McCleary, Tony Carnahan, Justin Teisberg, Dehrich Chya, Patrick Saltonstall, Amanda Lancaster, Leda Ferranti, Jimmie Christensen, Dorsey Burger, Jeff Selenger, Randy Gravatt, Tut Funtevilla, John and Christina Stark, Glyndaril White, and Marcella Amodo-White.

My gratitude also goes to Shannon Jamieson Vazquez, who read an early draft of the manuscript and offered helpful feedback on content, narrative arc, and structure. Joeth Zucco, a grizzly bear fan who had worked at both Yellowstone and Katmai National Parks, also provided some great edits, insights, and suggestions.

I owe a special thanks to Peter Mangolds and Jennifer Smith for allowing me to use many of their amazing bear photos that truly capture the beauty, mystery, and majesty of *Ursus arctos*. Thanks also goes to Doug and Lynne Seus, Randy Gravatt, and Nils Pedersen for their wonderful images.Mary Oliver's poetry served as a daily source of inspiration. Drained after a long day of research or writing, I'd read her poetry to reconnect with the magic and beauty of the natural world and fill up my creative well for the next day's work.

Most importantly, my immense gratitude goes to my wife, Meaghan, for her support (and for accompanying me on numerous bear-viewing adventures); and to my family—Mom, Dad, Kristine and Ola Johansson, Sean and Corie Grange; my niece, Lauren; and my nephews Bjorn, Finn, Hunter, and Taylor—along with the extended Morse and Grange clans.

Lastly, thank you to everyone around the world who loves bears and is working for them and their habitat.

Selected Bibliography

Books

Breiter, Matthias. *The Bears of Katmai.* Portland, OR: Graphic Arts Center Publishing, 1999.

Chadwick, Douglas. "Grizzly Cornered." *National Geographic*, July 2001.

Chya, Dehrich, and Amy F. Steffian. *Unigkuat: Kodiak Alutiiq Legends.* Kodiak, AK. Alutiiq Museum and Archaeological Repository, 2021.

Curwood, James Oliver. *The Grizzly King.* New York: Doubleday, Page & Co., 1916.

Craighead, Frank C. *Track of the Grizzly.* San Francisco, CA. Sierra Club Books, 1979.

Dodge, Harry B. *Kodiak Island and Its Bears.* Anchorage, AK: Great Northwest Publishing, 2004.

Fair, Jeff, and Larry Aumiller. *In Wild Trust.* Fairbanks: University of Alaska Press, 2017.

Fitz, Michael. *The Bears of Brooks Falls.* Taftsville, VT: Countryman Press, 2021.

Gunther, Kerry A., and Frank T. van Manen. *Yellowstone Grizzly Bears.* Yellowstone National Park: Yellowstone Forever, 2017.

Herrero, Stephen. *Bear Attacks: Their Causes and Avoidance.* 3rd ed. Essex, CT: Lyons Press, 2018.

Leopold, Aldo. *A Sand County Almanac.* New York: Oxford University Press, 1949.

Masterson, Linda. *Living with Bears Handbook.* Bellvue, CO: PixyJackPress, 2006.

Murie, Olaus J. *A Field Guide to Animal Tracks.* Boston, MA: Houghton Mifflin Company, 1954.

Murray, John. *The Great Bear.* Anchorage: Alaska Northwest Books, 1992.

Ouinth, Stefan. *Kodiak Alaska: Island of the Great Bear.* Anchorage, AK: Todd Communications, 2013.

Reardon, Jim. "The Kodiak Bear War." *Outdoor Life*, August 1964.

Rhode, Cecil E. "When Giant Alaskan Bears Go Fishing." *National Geographic*, August 1954.

Shepard, Paul, and Barry Sanders. *The Sacred Paw.* New York: Arcana, 1985.

Simpson, Sherry. *Dominion of Bears.* Lawrence: University Press of Kansas, 2013.

Van Daele, Larry. *The History of Bears on the Kodiak Archipelago.* Anchorage: Alaska Natural History Association, 2003.

Documentaries and Films

Annaud, Jean-Jacques. *The Bear.* Shout Factory Films, 1989.

Furbus, John. *Growing Up Grizzly.* Animal Planet, 2002.

Herzog, Werner. *Grizzly Man.* Lionsgate, 2008.

Lancaster, Bart. *Kodiak Island Brown Bears: Rohrer Bear Camp.* The Guide's Eye, 2010.

Anderson, Casey. *Project Kodiak.* National Geographic Society, 2009.

Organizations

Bearwise: www.bearwise.org

Bearsmart: www.bearsmart.com

Bristol Bay Defense Fund: stoppebbleminenow.org

Explore: www.explore.org

Friends of McNeil River: www.friendsofmcneilriver.org

Grand Teton National Park Foundation: www.gtnpf.org

Grizzly & Wolf Discovery Center: www.grizzlydiscoveryctr.org

Interagency Grizzly Bear Committee: www.igbconline.org

International Human-Bear Conflicts Workshop: www.humanbearconflicts.org

Katmai Conservancy: katmaiconservancy.org

Kodiak Brown Bear Trust: www.kodiakbrownbeartrust.org

Vital Ground Foundation: www.vitalground.org

Washington State University Bear Center: www.bearcenter.wsu.edu

Wind River Bear Institute: www.beardogs.org

Yellowstone Forever: www.yellowstone.org

About the Author

Kevin Grange is a firefighter paramedic in Jackson Hole, Wyoming. He is the award-winning author of *Wild Rescues: A Paramedic's Extreme Adventures in Yosemite, Yellowstone, and Grand Teton*; *Lights and Sirens: The Education of a Paramedic*; and *Beneath Blossom Rain: Discovering Bhutan on the Toughest Trek in the World*. He has written for *National Parks, Backpacker, Utne Reader, Yoga Journal*, and the *Orange County Register*. He has worked as a park ranger and paramedic at Yellowstone, Yosemite, and Grand Teton National Parks.